CYBERNETICS

or control and communication in the animal and the machine

NORBERT WIENER

PROFESSOR OF MATHEMATICS
THE MASSACHUSETTS INSTITUTE OF TECHNOLOGY

second edition

Martino Publishing
Mansfield Centre, CT
2013

Martino Publishing
P.O. Box 373,
Mansfield Centre, CT 06250 USA

ISBN   978-1-61427-502-2

© *2013 Martino Publishing*

Cover design by T. Matarazzo

Printed in the United States of America On 100% Acid-Free Paper

CYBERNETICS

or control and communication
in the animal and the machine

NORBERT WIENER

PROFESSOR OF MATHEMATICS
THE MASSACHUSETTS INSTITUTE OF TECHNOLOGY

second edition

THE M.I.T. PRESS
Cambridge, Massachusetts

Copyright © 1948 and 1961 by the Massachusetts Institute of Technology

Library of Congress Catalog Number 61-13034

Printed in the United States of America

To ARTURO ROSENBLUETH
for many years my companion in science

Preface
to the Second Edition

When I wrote the first edition of *Cybernetics* some thirteen years ago, I did it under some serious handicaps which had the effect of piling up unfortunate typographical errors, together with a few errors of content. Now I believe the time has come to reconsider cybernetics, not merely as a program to be carried out at some period in the future, but as an existing science. I have therefore taken this opportunity to put the necessary corrections at the disposal of my readers and, at the same time, to present an amplification of the present status of the subject and of the new related modes of thought which have come into being since its first publication.

If a new scientific subject has real vitality, the center of interest in it must and should shift in the course of years. When I first wrote *Cybernetics*, the chief obstacles which I found in making my point were that the notions of statistical information and control theory were novel and perhaps even shocking to the established attitudes of the time. At present, they have become so familiar as a tool of the communication engineers and of the designers of automatic controls that the chief danger against which I must guard is that the book may seem trite and commonplace. The role of feedback both in engineering design and in biology has come to be well established. The role of information and the technique of measuring and transmitting information constitute a whole discipline for the engineer, for the physiologist, for the psychologist, and for the sociologist. The automata which the first edition of this book barely forecast have come into their own, and the related social dangers against which I warned, not only in this book, but also in its small popular companion *The Human Use of Human Beings*,[1] have risen well above the horizon.

[1] Wiener, N., *The Human Use of Human Beings; Cybernetics and Society*, Houghton Mifflin Company, Boston, 1950.

Thus it behooves the cyberneticist to move on to new fields and to transfer a large part of his attention to ideas which have arisen in the developments of the last decade. The simple linear feedbacks, the study of which was so important in awakening scientists to the role of cybernetic study, now are seen to be far less simple and far less linear than they appeared at first view. Indeed, in the early days of electric circuit theory, the mathematical resources for systematic treatment of circuit networks did not go beyond linear juxtapositions of resistances, capacities, and inductances. This meant that the entire subject could be adequately described in terms of the harmonic analysis of the messages transmitted, and of the impedances, admittances, and voltage ratios of the circuits through which the messages were passed.

Long before the publication of *Cybernetics*, it came to be realized that the study of non-linear circuits (such as we find in many amplifiers, in voltage limiters, in rectifiers, and the like) did not fit easily into this frame. Nevertheless, for want of a better methodology, many attempts were made to extend the linear notions of the older electrical engineering well beyond the point where the newer types of apparatus could be naturally expressed in terms of these.

When I came to M.I.T. around 1920, the general mode of putting the questions concerning non-linear apparatus was to look for a direct extension of the notion of impedance which would cover linear as well as non-linear systems. The result was that the study of non-linear electrical engineering was getting into a state comparable with that of the last stages of the Ptolemaic system of astronomy, in which epicycle was piled on epicycle, correction upon correction, until a vast patchwork structure ultimately broke down under its own weight.

Just as the Copernican system arose out of the wreck of the overstrained Ptolemaic system, with a simple and natural heliocentric description of the motions of the heavenly bodies instead of the complicated and unperspicuous Ptolemaic geocentric system, so the study of non-linear structures and systems, whether electric or mechanical, whether natural or artificial, has needed a fresh and independent point of commencement. I have tried to initiate a new approach in my book *Nonlinear Problems in Random Theory*.[1] It turns out that the overwhelming importance of a trigonometric analysis in the treatment of linear phenomena does not persist when we come to consider non-linear phenomena. There is a clear-cut mathematical reason for this. Electrical circuit phenomena, like many other

[1] Wiener, N., *Nonlinear Problems in Random Theory*, The Technology Press of M.I.T. and John Wiley & Sons, Inc., New York, 1958.

physical phenomena, are characterized by an invariance with respect to a shift of origin in time. A physical experiment which will have arrived at a certain stage by 2 o'clock if we started at noon, will have arrived at the same stage at 2:15 if we started at 12:15. Thus the laws of physics concern invariants of the translation group in time.

The trigonometric functions $\sin nt$ and $\cos nt$ show certain important invariants with respect to the same translation group. The general function

$$e^{i\omega t}$$

will go into the function

$$e^{i\omega(t+\tau)} = e^{i\omega\tau}e^{i\omega t}$$

of the same form under the translation which we obtain by adding τ to t. As a consequence,

$a \cos n(t+\tau) + b \sin n(t+\tau)$

$$= (a \cos n\tau + b \sin n\tau) \cos nt + (b \cos n\tau - a \sin n\tau) \sin nt$$

$$= a_1 \cos nt + b_1 \sin nt$$

In other words, the families of functions

$$Ae^{i\omega t}$$

and

$$A \cos \omega t + B \sin \omega t$$

are invariant under translation.

Now there are other families of functions which are invariant under translations. If I consider the so-called random walk in which the movement of a particle under any time interval has a distribution dependent only on the length of that time interval and independent of everything that has happened up to its initiation, the ensemble of random walks will also go into itself under the time translation.

In other words, the mere translational invariance of the trigonometric curves is a property shared by other sets of functions as well.

The property which is characteristic of the trigonometric functions in addition to these invariants is that

$$Ae^{i\omega t} + Be^{i\omega t} = (A + B)e^{i\omega t}$$

so that these functions form an extremely simple linear set. It will be noted that this property concerns linearity; that is, that we can reduce all oscillations of a given frequency to a linear combination of two. It is this specific property which creates the value of harmonic analysis in the treatment of the linear properties of electric circuits. The functions

$$e^{i\omega t}$$

are characters of the translation group and yield a linear representation of this group.

When, however, we deal with combinations of functions other than addition with constant coefficients—when for example we multiply two functions by one another—the simple trigonometric functions no longer show this elementary group property. On the other hand, the random functions such as appear in the random walk do have certain properties very suitable for the discussion of their non-linear combinations.

It is scarcely desirable for me to go into the detail of this work here, for it is mathematically rather complicated, and it is covered in my book *Nonlinear Problems in Random Theory*. The material in that book has already been put to considerable use in the discussion of specific non-linear problems, but much remains to be done in carrying out the program laid down there. What it amounts to in practice is that an appropriate test input for the study of non-linear systems is rather of the character of the Brownian motion than a set of trigonometric functions. This Brownian motion function in the case of electric circuits can be generated physically by the shot effect. This shot effect is a phenomenon of irregularity in electrical currents which arises from the fact that such currents are carried not as a continuous stream of electricity but as a sequence of indivisible and equal electrons. Thus electric currents are subject to statistical irregularities which are themselves of a certain uniform character and which can be amplified up to the point at which they constitute an appreciable random noise.

As I shall show in Chapter IX, this theory of random noise can be put into practical use not merely for the analysis of electrical circuits and other non-linear processes but for their synthesis as well.[1] The device which is used is the reduction of the output of a non-linear instrument with random input to a well-defined series of certain orthonormal functions which are closely related to the Hermite polynomials. The problem of the analysis of a non-linear circuit consists in the determination of the coefficients of these polynomials in certain parameters of the input by a process of averaging.

The description of this process is rather simple. In addition to the black box which represents an as yet unanalyzed non-linear system, I have certain bodies of known structure which I shall call white

---

[1] Here I am using the term "non-linear system" not to exclude linear systems but to include a larger category of systems. The analysis of non-linear systems by means of random noise is also applicable to linear systems and has been so used.

boxes representing the various terms in the expansion desired.[1]   I put the same random noise into the black box and into a given white box.   The coefficient of the white box in the development of the black box is given as an average of the product of their outputs. While this average is taken over the entire ensemble of shot-effect inputs, there is a theorem which allows us to replace this average in all but a set of cases of probability 0 by an average taken over time. To obtain this average, we need to have at our disposal a multiplying instrument by which we can get the product of the outputs of the black and the white box, as well as an averaging instrument, which we can base on the fact that the potential across a condenser is proportional to the quantity of electricity held in the condenser and hence to the time integral of the current flowing through it.

Not only is it possible to determine the coefficients of each white box constituting an additive part of the equivalent representation of the black box one by one, but it is also possible to determine these quantities simultaneously.   It is even possible by the use of appropriate feedback devices to make each one of the white boxes automatically adjust itself to the level corresponding to its coefficient in the development of the black box.   In this manner we are able to construct a multiple white box which, when it is properly connected to a black box and is subjected to the same random input, will automatically form itself into an operational equivalent of the black box even though its internal structure may be vastly different.

These operations of analysis, synthesis, and automatic self-adjustment of white boxes into the likeness of black boxes can be carried out by other methods which have been described by Professor Amar Bose[2] and by Professor Gabor.[3]   In all of them there is a use of some process of working in, or learning, by choosing appropriate

[1] The terms "black box" and "white box" are convenient and figurative expressions of not very well determined usage.   I shall understand by a black box a piece of apparatus, such as four-terminal networks with two input and two output terminals, which performs a definite operation on the present and past of the input potential, but for which we do not necessarily have any information of the structure by which this operation is performed.   On the other hand, a white box will be a similar network in which we have built in the relation between input and output potentials in accordance with a definite structural plan for securing a previously determined input-output relation.

[2] Bose, A. G., "Nonlinear System Characterization and Optimization," *IRE Transactions on Information Theory*, IT-5, 30–40 (1959) (Special supplement to *IRE Transactions*).

[3] Gabor, D., "Electronic Inventions and Their Impact on Civilization," *Inaugural Lecture*, March 3, 1959, Imperial College of Science and Technology, University of London, England.

inputs for the black and white boxes and comparing them; and in many of these processes, including the method of Professor Gabor, multiplication devices play an important role. While there are many approaches to the problem of multiplying two functions electrically, this task is not technically easy. On the one hand, a good multiplier must work over a large range of amplitudes. On the other hand, it must be so nearly instantaneous in its operation that it will be accurate up to high frequencies. Gabor claims for his multiplier a frequency range running to about 1000 cycles. In his inaugural dissertation for the chair of Professor of Electrical Engineering at the Imperial College of Science and Technology of the University of London, he does not state explicitly the amplitude range over which his method of multiplication is valid nor the degree of accuracy to be obtained. I am awaiting very eagerly an explicit statement of these properties so that we can give a good evaluation of the multiplier for use in other pieces of apparatus dependent on it.

All of these devices in which an apparatus assumes a specific structure or function on the basis of past experience lead to a very interesting new attitude both in engineering and in biology. In engineering, devices of similar character can be used not only to play games and perform other purposive acts but to do so with a continual improvement of performance on the basis of past experience. I shall discuss some of these possibilities in Chapter IX of this book. Biologically, we have at least an analogue to what is perhaps the central phenomenon of life. For heredity to be possible and for cells to multiply, it is necessary that the heredity-carrying components of a cell—the so-called genes—be able to construct other similar heredity-carrying structures in their own image. It is, therefore, very exciting for us to be in possession of a means by which engineering structures can produce other structures with a function similar to their own. I shall devote Chapter X to this, and in particular shall discuss how oscillating systems of a given frequency can reduce other oscillating systems to the same frequency.

It is often stated that the production of any specific kind of molecule in the image of existing ones has an analogy to the use of templates in engineering whereby we can use a functional element of a machine as the pattern on which another similar element is made. The image of the template is a static one, and there must be some process by which one gene molecule manufactures another. I give the tentative suggestion that frequencies, let us say the frequencies of molecular spectra, may be the pattern elements which carry the identity of biological substances; and the self-organization of genes may be a manifestation of the self-organization of frequencies which I shall discuss later.

I have already spoken of learning machines in a general way. I shall devote a chapter to a more detailed discussion of these machines and potentialities and some of the problems of their use. Here I wish to make a few comments of a general nature.

As will be seen in Chapter I, the notion of learning machines is as old as cybernetics itself. In the anti-aircraft predictors which I described, the linear characteristics of the predictor which is used at any given time depend on a long-time acquaintance with the statistics of the ensemble of time series which we desire to predict. While a knowledge of these characteristics can be worked out mathematically in accordance with the principles which I have given there, it is perfectly possible to devise a computer which will work up these statistics and develop the short-time characteristics of the predictor on the basis of an experience which is already observed by the same machine as is used for prediction and which is worked up automatically. This can go far beyond the purely linear predictor. In various papers by Kallianpur, Masani, Akutowicz, and myself,[1] we have developed a theory of non-linear prediction which can at least conceivably be mechanized in a similar manner with the use of long-time observations to give the statistical basis for short-time prediction.

The theory of linear prediction and of non-linear prediction both involve some criteria of the goodness of fit of the prediction. The simplest criterion, although by no means the only usable one, is that of minimizing the mean square of the error. This is used in a particular form in connection with the functionals of the Brownian motion which I employ for the construction of non-linear apparatus, inasmuch as the various terms of my development have certain orthogonality properties. These ensure that the partial sum of a finite number of these terms is the best simulation of the apparatus to be imitated, which can be made by the employment of these terms if the mean square criterion of error is to be maintained. The work of Gabor also depends upon mean square criterion of error, but in a more general way, applicable to time series obtained by experience.

The notion of learning machines can be extended far beyond its

---

[1] Wiener, N., and P. Masani, "The Prediction Theory of Multivariate Stochastic Processes," Part I, *Acta Mathematica*, **98**, 111–150 (1957); Part II, *ibid.*, **99**, 93–137 (1958). Also Wiener, N., and E. J. Akutowicz, "The Definition and Ergodic Properties of the Stochastic Adjoint of a Unitary Transformation," *Rendiconti del Circolo Matematico di Palermo*, Ser. II, **VI**, 205–217 (1957).

employment for predictors, filters, and other similar apparatus. It is particularly important for the study and construction of machines which play a competitive game like checkers. Here the vital work has been done by Samuel[1] and Watanabe[2] at the laboratories of the International Business Machines Corporation. As in the case of filters and predictors, certain functions of the time series are developed in terms of which a much larger class of functions can be expanded. These functions can have numerical evaluations of the significant quantities on which the successful playing of a game depends. For example, they comprise the number of pieces on both sides, the total command of these pieces, their mobility, and so forth. At the beginning of the employment of the machine, these various considerations are given tentative weightings, and the machine chooses that admissible move for which the total weighting will have a maximum value. Up to this point, the machine has worked with a rigid program and has not been a learning machine.

However, at times the machine assumes a different task. It tries to expand that function which is 1 for won games, 0 for lost games, and perhaps $\frac{1}{2}$ for drawn games in terms of the various functions expressing the considerations of which the machine is able to take cognizance. In this way, it redetermines the weightings of these considerations so as to be able to play a more sophisticated game. I shall discuss some of the properties of these machines in Chapter IX, but here I must point out that they have been sufficiently successful for the machine to be able to defeat its programmer in from 10 to 20 hours of learning and working in. I also wish to mention in that chapter some of the work that has been done on similar machines devised for proving geometrical theorems and for simulating, to a limited extent, the logic of induction.

All of this work is a part of the theory and practice of the programming of programming, which has been extensively studied in the Electronic Systems Laboratory of the Massachusetts Institute of Technology. Here it has been found out that unless some such learning device is employed, the programming of a rigidly patterned machine is itself a very difficult task and that there is an urgent need for devices to program this programming.

Now that the concept of learning machines is applicable to those machines which we have made ourselves, it is also relevant to those

[1] Samuel, A. L., "Some Studies in Machine Learning, Using the Game of Checkers," *IBM Journal of Research and Development*, 3, 210–229 (1959).

[2] Watanabe, S., "Information Theoretical Analysis of Multivariate Correlation," *IBM Journal of Research and Development*, 4, 66–82 (1960).

living machines which we call animals, so that we have the possibility
of throwing a new light on biological cybernetics.   Here I wish to
single out, among a variety of current investigations, a book by the
Stanley-Jones on the Kybernetics (notice the spelling) of living
systems.[1]   In this book they devote a great deal of attention to those
feedbacks which maintain the working level of the nervous system as
well as those other feedbacks which respond to special stimuli.
Since the combination of the level of the system with the particular
responses is to a considerable extent multiplicative, it is also non-
linear and involves considerations of the sort we have already
brought out.   This field of activity is very much alive at present,
and I expect it to become much more alive in the near future.

The methods of memory machines and of machines that multiply
themselves which I have so far given are largely, although not
entirely, those which depend on apparatus of a high degree of specificity, or of what I may call blueprint apparatus.   The physiological
aspects of the same process must conform more to the peculiar
techniques of living organisms in which blueprints are replaced by a
less specific process, but one in which the system organizes itself.
Chapter X of this book is devoted to a sample of a self-organizing
process, namely, that by which narrow, highly specific frequencies
are formed in brain waves.   It is, therefore, largely the physiological
counterpart of the previous chapter, in which I am discussing similar
processes on more of a blueprint basis.   This existence of sharp
frequencies in brain waves and the theories which I gave to explain
how they are originated, what they can do, and what medical use
may be made of them represent in my mind an important and new
break-through in physiology.   Similar ideas can be used in many
other places in physiology and can make a real contribution to the
study of the fundamentals of life phenomena.   In this field, what I
am giving is more a program than work already achieved, but it is a
program for which I have great hopes.

It has not been my intention, either in the first edition or in the
present one, to make this book a compendium of all that has been
done in cybernetics.   Neither my interests nor my abilities lie that
way.   My intention is to express and to amplify my ideas on this
subject, and to display some of the ideas and philosophical reflections
which led me in the beginning to enter upon this field, and which
have continued to interest me in its development.   Thus it is an
intensely personal book, devoting much space to those developments

[1] Stanley-Jones, D., and K. Stanley-Jones, *Kybernetics of Natural Systems, A Study
in Patterns of Control*, Pergamon Press, London, 1960.

in which I myself have been interested, and relatively little to those in which I have not worked myself.

I have had valuable help from many quarters in revising this book. I must acknowledge in particular the cooperation of Miss Constance D. Boyd of The M.I.T. Press, Dr. Shikao Ikehara of the Tokyo Institute of Technology, Dr. Y. W. Lee of the Electrical Engineering Department of M.I.T., and Dr. Gordon Raisbeck of the Bell Telephone Laboratories. Also, in the writing down of my new chapters, and particularly in the computations of Chapter X, in which I have considered the case of self-organizing systems which manifest themselves in the study of the electroencephalogram, I wish to mention the aid which I received from my students, John C. Kotelly and Charles E. Robinson, and especially the contribution of Dr. John S. Barlow of the Massachusetts General Hospital. The indexing was done by James W. Davis.

Without the meticulous care and devotion of all of these I would not have had either the courage or the accuracy to turn out a new and corrected edition.

NORBERT WIENER

Cambridge, Massachusetts,
March, 1961

CONTENTS

PART I

ORIGINAL EDITION

1948

Introduction

This book represents the outcome, after more than a decade, of a program of work undertaken jointly with Dr. Arturo Rosenblueth, then of the Harvard Medical School and now of the Instituto Nacional de Cardiología of Mexico. In those days, Dr. Rosenblueth, who was the colleague and collaborator of the late Dr. Walter B. Cannon, conducted a monthly series of discussion meetings on scientific method. The participants were mostly young scientists at the Harvard Medical School, and we would gather for dinner about a round table in Vanderbilt Hall. The conversation was lively and unrestrained. It was not a place where it was either encouraged or made possible for anyone to stand on his dignity. After the meal, somebody—either one of our group or an invited guest—would read a paper on some scientific topic, generally one in which questions of methodology were the first consideration, or at least a leading consideration. The speaker had to run the gauntlet of an acute criticism, good-natured but unsparing. It was a perfect catharsis for half-baked ideas, insufficient self-criticism, exaggerated self-confidence, and pomposity. Those who could not stand the gaff did not return, but among the former habitués of these meetings there is more than one of us who feels that they were an important and permanent contribution to our scientific unfolding.

Not all the participants were physicians or medical scientists. One of us, a very steady member, and a great help to our discussions, was Dr. Manuel Sandoval Vallarta, a Mexican like Dr. Rosenblueth and a Professor of Physics at the Massachusetts Institute of Technology, who had been among my very first students when I came to the Institute after World War I. Dr. Vallarta used to bring some of his M.I.T. colleagues along to these discussion meetings, and it was at one of these that I first met Dr. Rosenblueth. I had been interested in the scientific method for a long time and had, in fact, been a participant in Josiah Royce's Harvard seminar on the subject

1

in 1911–1913. Moreover, it was felt that it was essential to have someone present who could examine mathematical questions critically. I thus became an active member of the group until Dr. Rosenblueth's call to Mexico in 1944 and the general confusion of the war ended the series of meetings.

For many years Dr. Rosenblueth and I had shared the conviction that the most fruitful areas for the growth of the sciences were those which had been neglected as a no-man's land between the various established fields. Since Leibniz there has perhaps been no man who has had a full command of all the intellectual activity of his day. Since that time, science has been increasingly the task of specialists, in fields which show a tendency to grow progressively narrower. A century ago there may have been no Leibniz, but there was a Gauss, a Faraday, and a Darwin. Today there are few scholars who can call themselves mathematicians or physicists or biologists without restriction. A man may be a topologist or an acoustician or a coleopterist. He will be filled with the jargon of his field, and will know all its literature and all its ramifications, but, more frequently than not, he will regard the next subject as something belonging to his colleague three doors down the corridor, and will consider any interest in it on his own part as an unwarrantable breach of privacy.

These specialized fields are continually growing and invading new territory. The result is like what occurred when the Oregon country was being invaded simultaneously by the United States settlers, the British, the Mexicans, and the Russians—an inextricable tangle of exploration, nomenclature, and laws. There are fields of scientific work, as we shall see in the body of this book, which have been explored from the different sides of pure mathematics, statistics, electrical engineering, and neurophysiology; in which every single notion receives a separate name from each group, and in which important work has been triplicated or quadruplicated, while still other important work is delayed by the unavailability in one field of results that may have already become classical in the next field.

It is these boundary regions of science which offer the richest opportunities to the qualified investigator. They are at the same time the most refractory to the accepted techniques of mass attack and the division of labor. If the difficulty of a physiological problem is mathematical in essence, ten physiologists ignorant of mathematics will get precisely as far as one physiologist ignorant of mathematics, and no further. If a physiologist who knows no mathematics works together with a mathematician who knows no physiology, the one will be unable to state his problem in terms that the other can manipulate, and the second will be unable to put the answers in any form that the first can understand. Dr. Rosenblueth has always insisted that a proper exploration of these blank spaces on the map of science could only be made by a team of scientists, each a specialist in his own field but each possessing a thoroughly sound and trained acquaintance with the fields of his neighbors; all in the habit of working together, of knowing one another's intellectual customs, and of recognizing the significance of a colleague's new suggestion before it has taken on a full formal expression. The mathematician need not have the skill to conduct a physiological experiment, but he must have the skill to understand one, to criticize one, and to suggest one. The physiologist need not be able to prove a certain mathematical theorem, but he must be able to grasp its physiological significance and to tell the mathematician for what he should look. We had dreamed for years of an institution of independent scientists, working together in one of these backwoods of science, not as subordinates of some great executive officer, but joined by the desire, indeed by the spiritual necessity, to understand the region as a whole, and to lend one another the strength of that understanding.

We had agreed on these matters long before we had chosen the field of our joint investigations and our respective parts in them. The deciding factor in this new step was the war. I had known for a considerable time that if a national emergency should come, my function in it would be determined largely by two things: my close contact with the program of computing machines developed by Dr. Vannevar Bush, and my own joint work with Dr. Yuk Wing Lee on the design of electric networks. In fact, both proved important. In the summer of 1940, I turned a large part of my attention to the development of computing machines for the solution of partial differential equations. I had long been interested in these and had convinced myself that their chief problem, as contrasted with the ordinary differential equations so well treated by Dr. Bush on his differential analyzer, was that of the representation of functions of more than one variable. I had also become convinced that the process of scanning, as employed in television, gave the answer to that question and, in fact, that television was destined to be more useful to engineering by the introduction of such new techniques than as an independent industry.

It was clear that any scanning process must vastly increase the number of data dealt with as compared with the number of data in a problem of ordinary differential equations. To accomplish reasonable results in a reasonable time, it thus became necessary to push the speed of the elementary processes to the maximum, and to avoid interrupting the stream of these processes by steps of an essentially slower nature. It also became necessary to perform the individual processes with so high a degree of accuracy that the enormous repetition of the elementary processes should not bring about a cumulative error so great as to swamp all accuracy. Thus the following requirements were suggested:

1. That the central adding and multiplying apparatus of the computing machine should be numerical, as in an ordinary adding machine, rather than on a basis of measurement, as in the Bush differential analyzer.

2. That these mechanisms, which are essentially switching devices, should depend on electronic tubes rather than on gears or mechanical relays, in order to secure quicker action.

3. That, in accordance with the policy adopted in some existing apparatus of the Bell Telephone Laboratories, it would probably be more economical in apparatus to adopt the scale of two for addition and multiplication, rather than the scale of ten.

4. That the entire sequence of operations be laid out on the machine itself so that there should be no human intervention from the time the data were entered until the final results should be taken off, and that all logical decisions necessary for this should be built into the machine itself.

5. That the machine contain an apparatus for the storage of data which should record them quickly, hold them firmly until erasure, read them quickly, erase them quickly, and then be immediately available for the storage of new material.

These recommendations, together with tentative suggestions for the means of realizing them, were sent in to Dr. Vannevar Bush for their possible use in a war. At that stage of the preparations for war, they did not seem to have sufficiently high priority to make immediate work on them worth while. Nevertheless, they all represent ideas which have been incorporated into the modern ultra-rapid computing machine. These notions were all very much in the spirit of the thought of the time, and I do not for a moment wish to claim anything like the sole responsibility for their introduction. Nevertheless, they have proved useful, and it is my hope that my memorandum had some effect in popularizing them among engineers. At any rate, as we shall see in the body of the book, they are all ideas which are of interest in connection with the study of the nervous system.

This work was thus laid on the table, and, although it has not proved to be fruitless, it led to no immediate project by Dr. Rosenblueth and myself. Our actual collaboration resulted from another project, which was likewise undertaken for the purposes of the last war. At the beginning of the war, the German prestige in aviation and the defensive position of England turned the attention of many scientists to the improvement of anti-aircraft artillery. Even before the war, it had become clear that the speed of the airplane had rendered obsolete all classical methods of the direction of fire, and that it was necessary to build into the control apparatus all the computations necessary. These were rendered much more difficult by the fact that, unlike all previously encountered targets, an airplane has a velocity which is a very appreciable part of the velocity of the missile used to bring it down. Accordingly, it is exceedingly important to shoot the missile, not at the target, but in such a way that missile and target may come together in space at some time in the future. We must hence find some method of predicting the future position of the plane.

The simplest method is to extrapolate the present course of the plane along a straight line. This has much to recommend it. The more a plane doubles and curves in flight, the less is its effective velocity, the less time it has to accomplish a mission, and the longer it remains in a dangerous region. Other things being equal, a plane will fly as straight a course as possible. However, by the time the first shell has burst, other things are *not* equal, and the pilot will probably zigzag, stunt, or in some other way take evasive action.

If this action were completely at the disposal of the pilot, and the pilot were to make the sort of intelligent use of his chances that we anticipate in a good poker player, for example, he has so much opportunity to modify his expected position before the arrival of a shell that we should not reckon the chances of hitting him to be very good, except perhaps in the case of a very wasteful barrage fire. On the other hand, the pilot does *not* have a completely free chance to maneuver at his will. For one thing, he is in a plane going at an exceedingly high speed, and any too sudden deviation from his course will produce an acceleration that will render him unconscious and may disintegrate the plane. Then too, he can control the plane only by moving his control surfaces, and the new regimen of flow that is established takes some small time to develop. Even when it is fully developed, it merely changes the acceleration of the plane, and this change of acceleration must be converted, first into change of velocity and then into change of position, before it is finally effective.

Moreover, an aviator under the strain of combat conditions is scarcely in a mood to engage in any very complicated and untrammeled voluntary behavior, and is quite likely to follow out the pattern of activity in which he has been trained.

All this made an investigation of the problem of the curvilinear prediction of flight worth while, whether the results should prove favorable or unfavorable for the actual use of a control apparatus involving such curvilinear prediction. To predict the future of a curve is to carry out a certain operation on its past. The true prediction operator cannot be realized by any constructible apparatus; but there are certain operators which bear it a certain resemblance and are, in fact, realizable by apparatus which we can build. I suggested to Professor Samuel Caldwell of the Massachusetts Institute of Technology that these operators seemed worth trying, and he immediately suggested that we try them out on Dr. Bush's differential analyzer, using this as a ready-made model of the desired fire-control apparatus. We did so, with results which will be discussed in the body of this book. At any rate, I found myself engaged in a war project, in which Mr. Julian H. Bigelow and myself were partners in the investigation of the theory of prediction and of the construction of apparatus to embody these theories.

It will be seen that for the second time I had become engaged in the study of a mechanico-electrical system which was designed to usurp a specifically human function—in the first case, the execution of a complicated pattern of computation, and in the second, the forecasting of the future. In this second case, we should not avoid the discussion of the performance of certain human functions. In some fire-control apparatus, it is true, the original impulse to point comes in directly by radar, but in the more usual case, there is a human gun-pointer or a gun-trainer or both coupled into the fire-control system, and acting as an essential part of it. It is essential to know their characteristics, in order to incorporate them mathematically into the machines they control. Moreover, their target, the plane, is also humanly controlled, and it is desirable to know its performance characteristics.

Mr. Bigelow and I came to the conclusion that an extremely important factor in voluntary activity is what the control engineers term *feedback*. I shall discuss this in considerable detail in the appropriate chapters. It is enough to say here that when we desire a motion to follow a given pattern the difference between this pattern and the actually performed motion is used as a new input to cause the part regulated to move in such a way as to bring its motion closer to that given by the pattern. For example, one form of steering engine of a ship carries the reading of the wheel to an offset from the tiller, which so regulates the valves of the steering engine as to move the tiller in such a way as to turn these valves off. Thus the tiller turns so as to bring the other end of the valve-regulating offset amidships, and in that way registers the angular position of the wheel as the angular position of the tiller. Clearly, any friction or other delaying force which hampers the motion of the tiller will increase the admission of steam to the valves on one side and will decrease it on the other, in such a way as to increase the torque tending to bring the tiller to the desired position. Thus the feedback system tends to make the performance of the steering engine relatively independent of the load.

On the other hand, under certain conditions of delay, etc., a feedback that is too brusque will make the rudder overshoot, and will be followed by a feedback in the other direction, which makes the rudder overshoot still more, until the steering mechanism goes into a wild oscillation or *hunting*, and breaks down completely. In a book such as that by MacColl,[1] we find a very precise discussion of feedback, the conditions under which it is advantageous, and the conditions under which it breaks down. It is a phenomenon which we understand very thoroughly from a quantitative point of view.

Now, suppose that I pick up a lead pencil. To do this, I have to move certain muscles. However, for all of us but a few expert anatomists, we do not know what these muscles are; and even among the anatomists, there are few, if any, who can perform the act by a conscious willing in succession of the contraction of each muscle concerned. On the contrary, what we will is *to pick the pencil up*. Once we have determined on this, our motion proceeds in such a way that we may say roughly that the amount by which the pencil is not yet picked up is decreased at each stage. This part of the action is not in full consciousness.

To perform an action in such a manner, there must be a report to the nervous system, conscious or unconscious, of the amount by which we have failed to pick up the pencil at each instant. If we have our eye on the pencil, this report may be visual, at least in part, but it is more generally kinesthetic, or, to use a term now in vogue, proprioceptive. If the proprioceptive sensations are wanting and we do not replace them by a visual or other substitute, we are unable to perform the act of picking up the pencil, and find ourselves

[1] MacColl, L. A., *Fundamental Theory of Servomechanisms*, Van Nostrand, New York, 1946.

in a state of what is known as *ataxia*. An ataxia of this type is familiar in the form of syphilis of the central nervous system known as *tabes dorsalis*, where the kinesthetic sense conveyed by the spinal nerves is more or less destroyed.

However, an excessive feedback is likely to be as serious a handicap to organized activity as a defective feedback. In view of this possibility, Mr. Bigelow and myself approached Dr. Rosenblueth with a very specific question. Is there any pathological condition in which the patient, in trying to perform some voluntary act like picking up a pencil, overshoots the mark, and goes into an uncontrollable oscillation? Dr. Rosenblueth immediately answered us that there is such a well-known condition, that it is called purpose tremor, and that it is often associated with injury to the cerebellum.

We thus found a most significant confirmation of our hypothesis concerning the nature of at least some voluntary activity. It will be noted that our point of view considerably transcended that current among neurophysiologists. The central nervous system no longer appears as a self-contained organ, receiving inputs from the senses and discharging into the muscles. On the contrary, some of its most characteristic activities are explicable only as circular processes, emerging from the nervous system into the muscles, and re-entering the nervous system through the sense organs, whether they be proprioceptors or organs of the special senses. This seemed to us to mark a new step in the study of that part of neurophysiology which concerns not solely the elementary processes of nerves and synapses but the performance of the nervous system as an integrated whole.

The three of us felt that this new point of view merited a paper, which we wrote up and published.[1] Dr. Rosenblueth and I foresaw that this paper could be only a statement of program for a large body of experimental work, and we decided that if we could ever bring our plan for an interscientific institute to fruition, this topic would furnish an almost ideal center for our activity.

On the communication engineering plane, it had already become clear to Mr. Bigelow and myself that the problems of control engineering and of communication engineering were inseparable, and that they centered not around the technique of electrical engineering but around the much more fundamental notion of the message, whether this should be transmitted by electrical, mechanical, or nervous means. The message is a discrete or continuous sequence of measurable events distributed in time—precisely what is called a time series

[1] Rosenblueth, A., N. Wiener, and J. Bigelow, "Behavior, Purpose, and Teleology," *Philosophy of Science*, **10**, 18–24 (1943).

by the statisticians. The prediction of the future of a message is done by some sort of operator on its past, whether this operator is realized by a scheme of mathematical computation, or by a mechanical or electrical apparatus. In this connection, we found that the ideal prediction mechanisms which we had at first contemplated were beset by two types of error, of a roughly antagonistic nature. While the prediction apparatus which we at first designed could be made to anticipate an extremely smooth curve to any desired degree of approximation, this refinement of behavior was always attained at the cost of an increasing sensitivity. The better the apparatus was for smooth waves, the more it would be set into oscillation by small departures from smoothness, and the longer it would be before such oscillations would die out. Thus the good prediction of a smooth wave seemed to require a more delicate and sensitive apparatus than the best possible prediction of a rough curve, and the choice of the particular apparatus to be used in a specific case was dependent on the statistical nature of the phenomenon to be predicted. This interacting pair of types of error seemed to have something in common with the contrasting problems of the measure of position and of momentum to be found in the Heisenberg quantum mechanics, as described according to his Principle of Uncertainty.

Once we had clearly grasped that the solution of the problem of optimum prediction was only to be obtained by an appeal to the statistics of the time series to be predicted, it was not difficult to make what had originally seemed to be a difficulty in the theory of prediction into what was actually an efficient tool for solving the problem of prediction. Assuming the statistics of a time series, it became possible to derive an explicit expression for the mean square error of prediction by a given technique and for a given lead. Once we had this, we could translate the problem of optimum prediction to the determination of a specific operator which should reduce to a minimum a specific positive quantity dependent on this operator. Minimization problems of this type belong to a recognized branch of mathematics, the calculus of variations, and this branch has a recognized technique. With the aid of this technique, we were able to obtain an explicit best solution of the problem of predicting the future of a time series, given its statistical nature, and even further, to achieve a physical realization of this solution by a constructible apparatus.

Once we had done this, at least one problem of engineering design took on a completely new aspect. In general, engineering design has been held to be an art rather than a science. By reducing a problem of this sort to a minimization principle, we had established the subject on a far more scientific basis. It occurred to us that this was not an isolated case, but that there was a whole region of engineering work in which similar design problems could be solved by the methods of the calculus of variations.

We attacked and solved other similar problems by the same methods. Among these was the problem of the design of wave filters. We often find a message contaminated by extraneous disturbances which we call *background noise*. We then face the problem of restoring the original message, or the message under a given lead, or the message modified by a given lag, by an operator applied to the corrupted message. The optimum design of this operator and of the apparatus by which it is realized depends on the statistical nature of the message and the noise, singly and jointly. We thus have replaced in the design of wave filters processes which were formerly of an empirical and rather haphazard nature by processes with a thorough scientific justification.

In doing this, we have made of communication engineering design a statistical science, a branch of statistical mechanics. The notion of statistical mechanics has indeed been encroaching on every branch of science for more than a century. We shall see that this dominance of statistical mechanics in modern physics has a very vital significance for the interpretation of the nature of time. In the case of communication engineering, however, the significance of the statistical element is immediately apparent. The transmission of information is impossible save as a transmission of alternatives. If only one contingency is to be transmitted, then it may be sent most efficiently and with the least trouble by sending no message at all. The telegraph and the telephone can perform their function only if the messages they transmit are continually varied in a manner not completely determined by their past, and can be designed effectively only if the variation of these messages conforms to some sort of statistical regularity.

To cover this aspect of communication engineering, we had to develop a statistical theory of the *amount of information*, in which the unit amount of information was that transmitted as a single decision between equally probable alternatives. This idea occurred at about the same time to several writers, among them the statistician R. A. Fisher, Dr. Shannon of the Bell Telephone Laboratories, and the author. Fisher's motive in studying this subject is to be found in classical statistical theory; that of Shannon in the problem of coding information; and that of the author in the problem of noise and message in electrical filters.   Let it be remarked parenthetically that some of my speculations in this direction attach themselves to the earlier work of Kolmogoroff[1] in Russia, although a considerable part of my work was done before my attention was called to the work of the Russian school.

The notion of the amount of information attaches itself very naturally to a classical notion in statistical mechanics: that of *entropy*.   Just as the amount of information in a system is a measure of its degree of organization, so the entropy of a system is a measure of its degree of disorganization; and the one is simply the negative of the other.   This point of view leads us to a number of considerations concerning the second law of thermodynamics, and to a study of the possibility of the so-called Maxwell demons.   Such questions arise independently in the study of enzymes and other catalysts, and their study is essential for the proper understanding of such fundamental phenomena of living matter as metabolism and reproduction.   The third fundamental phenomenon of life, that of irritability, belongs to the domain of communication theory and falls under the group of ideas we have just been discussing.[2]

Thus, as far back as four years ago, the group of scientists about Dr. Rosenblueth and myself had already become aware of the essential unity of the set of problems centering about communication, control, and statistical mechanics, whether in the machine or in living tissue.   On the other hand, we were seriously hampered by the lack of unity of the literature concerning these problems, and by the absence of any common terminology, or even of a single name for the field.   After much consideration, we have come to the conclusion that all the existing terminology has too heavy a bias to one side or another to serve the future development of the field as well as it should; and as happens so often to scientists, we have been forced to coin at least one artificial neo-Greek expression to fill the gap.   We have decided to call the entire field of control and communication theory, whether in the machine or in the animal, by the name *Cybernetics*, which we form from the Greek χυβερνήτης or *steersman*. In choosing this term, we wish to recognize that the first significant paper on feedback mechanisms is an article on governors, which was published by Clerk Maxwell in 1868,[3] and that *governor* is

[1] Kolmogoroff, A. N., "Interpolation und Extrapolation von stationären Zufälligen Folgen," *Bull. Acad. Sci. U.S.S.R.*, Ser. Math. **5**, 3–14 (1941).

[2] Schrödinger, Erwin, *What is Life?*, Cambridge University Press, Cambridge, England, 1945.

[3] Maxwell, J. C., *Proc. Roy. Soc. (London)*, **16**, 270–283, (1868).

derived from a Latin corruption of χυβερνήτης. We also wish to refer to the fact that the steering engines of a ship are indeed one of the earliest and best-developed forms of feedback mechanisms.

Although the term *cybernetics* does not date further back than the summer of 1947, we shall find it convenient to use in referring to earlier epochs of the development of the field. From 1942 or thereabouts, the development of the subject went ahead on several fronts. First, the ideas of the joint paper by Bigelow, Rosenblueth, and Wiener were disseminated by Dr. Rosenblueth at a meeting held in New York in 1942, under the auspices of the Josiah Macy Foundation, and devoted to problems of central inhibition in the nervous system. Among those present at that meeting was Dr. Warren McCulloch, of the Medical School of the University of Illinois, who had already been in touch with Dr. Rosenblueth and myself, and who was interested in the study of the organization of the cortex of the brain.

At this point there enters an element which occurs repeatedly in the history of cybernetics—the influence of mathematical logic. If I were to choose a patron saint for cybernetics out of the history of science, I should have to choose Leibniz. The philosophy of Leibniz centers about two closely related concepts—that of a universal symbolism and that of a calculus of reasoning. From these are descended the mathematical notation and the symbolic logic of the present day. Now, just as the calculus of arithmetic lends itself to a mechanization progressing through the abacus and the desk computing machine to the ultra-rapid computing machines of the present day, so the *calculus ratiocinator* of Leibniz contains the germs of the *machina ratiocinatrix*, the reasoning machine. Indeed, Leibniz himself, like his predecessor Pascal, was interested in the construction of computing machines in the metal. It is therefore not in the least surprising that the same intellectual impulse which has led to the development of mathematical logic has at the same time led to the ideal or actual mechanization of processes of thought.

A mathematical proof which we can follow is one which can be written in a finite number of symbols. These symbols, in fact, may make an appeal to the notion of infinity, but this appeal is one which we can sum up in a finite number of stages, as in the case of mathematical induction, where we prove a theorem depending on a parameter n for $n = 0$, and also prove that the case $n + 1$ follows from the case n, thus establishing the theorem for all positive values of n. Moreover, the rules of operation of our deductive mechanism must be finite in number, even though they may appear to be otherwise, through a reference to the concept of infinity, which can itself be stated in finite terms.   In short, it has become quite evident, both to the nominalists like Hilbert and to the intuitionists like Weyl, that the development of a mathematico-logical theory is subject to the same sort of restrictions as those that limit the performance of a computing machine.   As we shall see later, it is even possible to interpret in this way the paradoxes of Cantor and of Russell.

I am myself a former student of Russell and owe much to his influence.   Dr. Shannon took for his doctor's thesis at the Massachusetts Institute of Technology the application of the techniques of the classical Boolean algebra of classes to the study of switching systems in electrical engineering.   Turing, who is perhaps first among those who have studied the logical possibilities of the machine as an intellectual experiment, served the British government during the war as a worker in electronics, and is now in charge of the program which the National Physical Laboratory at Teddington has undertaken for the development of computing machines of the modern type.

Another young migrant from the field of mathematical logic to cybernetics is Walter Pitts.   He had been a student of Carnap at Chicago and had also been in contact with Professor Rashevsky and his school of biophysicists.   Let it be remarked in passing that this group has contributed much to directing the attention of the mathematically minded to the possibilities of the biological sciences, although it may seem to some of us that they are too dominated by problems of energy and potential and the methods of classical physics to do the best possible work in the study of systems like the nervous system, which are very far from being closed energetically.

Mr. Pitts had the good fortune to fall under McCulloch's influence, and the two began to work quite early on problems concerning the union of nerve fibers by synapses into systems with given over-all properties.   Independently of Shannon, they had used the technique of mathematical logic for the discussion of what were after all switching problems.   They added elements which were not prominent in Shannon's earlier work, although they are certainly suggested by the ideas of Turing: the use of the time as a parameter, the consideration of nets containing cycles, and of synaptic and other delays.[1]

In the summer of 1943, I met Dr. J. Lettvin of the Boston City Hospital, who was very much interested in matters concerning

[1] Turing, A. M., "On Computable Numbers, with an Application to the Entscheidungsproblem," *Proceedings of the London Mathematical Society*, Ser. 2, **42**, 230–265 (1936).

nervous mechanisms.   He was a close friend of Mr. Pitts, and made
me acquainted with his work.[1]   He induced Mr. Pitts to come out to
Boston, and to make the acquaintance of Dr. Rosenblueth and myself.   We welcomed him into our group.   Mr. Pitts came to the
Massachusetts Institute of Technology in the autumn of 1943, in
order to work with me and to strengthen his mathematical background for the study of the new science of cybernetics, which had by
that time been fairly born but not yet christened.

At that time Mr. Pitts was already thoroughly acquainted with
mathematical logic and neurophysiology, but had not had the chance
to make very many engineering contacts.   In particular, he was not
acquainted with Dr. Shannon's work, and he had not had much
experience of the possibilities of electronics.   He was very much
interested when I showed him examples of modern vacuum tubes and
explained to him that these were ideal means for realizing in the metal
the equivalents of his neuronic circuits and systems.   From that
time, it became clear to us that the ultra-rapid computing machine,
depending as it does on consecutive switching devices, must represent
almost an ideal model of the problems arising in the nervous system.
The all-or-none character of the discharge of the neurons is precisely
analogous to the single choice made in determining a digit on the
binary scale, which more than one of us had already contemplated as
the most satisfactory basis of computing-machine design.   The
synapse is nothing but a mechanism for determining whether a certain
combination of outputs from other selected elements will or will not
act as an adequate stimulus for the discharge of the next element,
and must have its precise analogue in the computing machine.   The
problem of interpreting the nature and varieties of memory in the
animal has its parallel in the problem of constructing artificial
memories for the machine.

At this time, the construction of computing machines had proved
to be more essential for the war effort than the first opinion of Dr.
Bush might have indicated, and was progressing at several centers
along lines not too different from those which my earlier report had
indicated.   Harvard, Aberdeen Proving Ground, and the University
of Pennsylvania were already constructing machines, and the
Institute for Advanced Study at Princeton and the Massachusetts
Institute of Technology were soon to enter the same field.   In this
program there was a gradual progress from the mechanical assembly
to the electrical assembly, from the scale of ten to the scale of two,

[1] McCulloch, W. S., and W. Pitts, "A logical calculus of the ideas immanent in
nervous activity," *Bull. Math. Biophys*, 5, 115–133 (1943).

from the mechanical relay to the electrical relay, from humanly directed operation to automatically directed operation; and in short, each new machine more than the last was in conformity with the memorandum I had sent Dr. Bush. There was a continual going and coming of those interested in these fields. We had an opportunity to communicate our ideas to our colleagues, in particular to Dr. Aiken of Harvard, Dr. von Neumann of the Institute for Advanced Study, and Dr. Goldstine of the Eniac and Edvac machines at the University of Pennsylvania. Everywhere we met with a sympathetic hearing, and the vocabulary of the engineers soon became contaminated with the terms of the neurophysiologist and the psychologist.

At this stage of the proceedings, Dr. von Neumann and myself felt it desirable to hold a joint meeting of all those interested in what we now call cybernetics, and this meeting took place at Princeton in the late winter of 1943–1944. Engineers, physiologists, and mathematicians were all represented. It was impossible to have Dr. Rosenblueth among us, as he had just accepted an invitation to act as Head of the laboratories of physiology of the Instituto Nacional de Cardiología in Mexico, but Dr. McCulloch and Dr. Lorente de Nó of the Rockefeller Institute represented the physiologists. Dr. Aiken was unable to be present; however, Dr. Goldstine was one of a group of several computing-machine designers who participated in the meeting, while Dr. von Neumann, Mr. Pitts, and myself were the mathematicians. The physiologists gave a joint presentation of cybernetic problems from their point of view; similarly, the computing-machine designers presented their methods and objectives. At the end of the meeting, it had become clear to all that there was a substantial common basis of ideas between the workers in the different fields, that people in each group could already use notions which had been better developed by the others, and that some attempt should be made to achieve a common vocabulary.

A considerable period before this, the war research group conducted by Dr. Warren Weaver had published a document, first secret and later restricted, covering the work of Mr. Bigelow and myself on predictors and wave filters. It was found that the conditions of anti-aircraft fire did not justify the design of special apparatus for curvilinear prediction, but the principles proved to be sound and practical, and have been used by the government for smoothing purposes, and in several fields of related work. In particular, the type of integral equation to which the calculus of variations problem reduces itself has been shown to emerge in wave-guide problems and

in many other problems of an applied mathematical interest.   Thus
in one way or another, the end of the war saw the ideas of prediction
theory and of the statistical approach to communication engineering
already familiar to a large part of the statisticians and communication engineers of the United States and Great Britain.   It also saw
my government document, now out of print, and a considerable
number of expository papers by Levinson,[1] Wallman, Daniell,
Phillips, and others written to fill the gap.   I myself have had a long
mathematical expository paper under way for several years to put
the work I have done on permanent record, but circumstances not
completely under my control have prevented its prompt publication.
Finally, after a joint meeting at the American Mathematical Society
and the Institute of Mathematical Statistics held in New York in
the spring of 1947, and devoted to the study of stochastic processes
from a point of view closely allied to cybernetics, I have passed on
what I have already written of my manuscript to Professor Doob of
the University of Illinois, to be developed in his notation and according to his ideas as a book for the Mathematical Surveys series of the
American Mathematical Society.   I had already developed part of
my work in a course of lectures in the mathematics department of
M.I.T. in the summer of 1945.   Since then, my old student and
collaborator,[2] Dr. Y. W. Lee, has returned from China.   He is
giving a course on the new methods for the design of wave filters and
similar apparatus in the M.I.T. electrical engineering department in
the fall of 1947, and has plans to work the material of these lectures
up into a book.   At the same time, the out-of-print government
document is to be reprinted.[3]

As I have said, Dr. Rosenblueth returned to Mexico about the
beginning of 1944.   In the spring of 1945, I received an invitation
from the Mexican Mathematical Society to participate in a meeting
to be held in Guadalajara that June.   This invitation was reinforced
by the Comision Instigadora y Coordinadora de la Investigación
Cientifica, under the leadership of Dr. Manuel Sandoval Vallarta, of
whom I have already spoken.   Dr. Rosenblueth invited me to share
some scientific research with him, and the Instituto Nacional de
Cardiología, under its director Dr. Ignacio Chávez, extended me its
hospitality.

I stayed some ten weeks in Mexico at that time.   Dr. Rosenblueth

[1] Levinson, N., *J. Math. and Physics*, **25**, 261–278; **26**, 110–119 (1947).
[2] Lee, Y. W., *J. Math. and Physics*, **11**, 261–278 (1932).
[3] Wiener, N., *Extrapolation, Interpolation, and Smoothing of Stationary Time Series*, Technology Press and Wiley, New York, 1949.

and I decided to continue a line of work which we had already discussed with Dr. Walter B. Cannon, who was also with Dr. Rosenblueth, on a visit which unfortunately proved to be his last. This work had to do with the relation between, on the one hand, the tonic, clonic, and phasic contractions in epilepsy and, on the other hand, the tonic spasm, beat, and fibrillation of the heart. We felt that heart muscle represented an irritable tissue as useful for the investigation of conduction mechanisms as nerve tissue, and furthermore, that the anastomoses and decussations of the heart-muscle fibers presented us with a simpler phenomenon than the problem of the nervous synapse. We were also deeply grateful to Dr. Chávez for his unquestioning hospitality, and, while it has never been the policy of the Instituto to restrict Dr. Rosenblueth to the investigation of the heart, we were grateful to have an opportunity to contribute to its principal purpose.

Our investigation took two directions: the study of phenomena of conductivity and latency in uniform conducting media of two or more dimensions, and the statistical study of the conducting properties of random nets of conducting fibers. The first led us to the rudiments of a theory of heart flutter, the latter to a certain possible understanding of fibrillation. Both lines of work were developed in a paper,[1] published by us, and, although in both cases our earlier results have shown the need of a considerable amount of revision and of supplementation, the work on flutter is being revised by Mr. Oliver G. Selfridge of the Massachusetts Institute of Technology, while the statistical technique used in the study of heart-muscle nets has been extended to the treatment of neuronal nets by Mr. Walter Pitts, now a Fellow of the John Simon Guggenheim Foundation. The experimental work is being carried on by Dr. Rosenblueth with the aid of Dr. F. García Ramos of the Instituto Nacional de Cardiología and the Mexican Army Medical School.

At the Guadalajara meeting of the Mexican Mathematical Society, Dr. Rosenblueth and I presented some of our results. We had already come to the conclusion that our earlier plans of collaboration had shown themselves to be practicable. We were fortunate enough to have a chance to present our results to a larger audience. In the spring of 1946, Dr. McCulloch had made arrangements with the Josiah Macy Foundation for the first of a series of meetings to be held in New York and to be devoted to the problems of feedback.

---

[1] Wiener, N., and A. Rosenblueth, "The Mathematical Formulation of the Problem of Conduction of Impulses in a Network of Connected Excitable Elements, Specifically in Cardiac Muscle," *Arch. Inst. Cardiol. Méx.*, **16**, 205–265 (1946).

These meetings have been conducted in the traditional Macy way, worked out most efficiently by Dr. Frank Fremont-Smith, who organized them on behalf of the Foundation. The idea has been to get together a group of modest size, not exceeding some twenty in number, of workers in various related fields, and to hold them together for two successive days in all-day series of informal papers, discussions, and meals together, until they had had the opportunity to thresh out their differences and to make progress in thinking along the same lines. The nucleus of our meetings has been the group that had assembled in Princeton in 1944, but Drs. McCulloch and Fremont-Smith have rightly seen the psychological and sociological implications of the subject, and have co-opted into the group a number of leading psychologists, sociologists, and anthropologists. The need of including psychologists had indeed been obvious from the beginning. He who studies the nervous system cannot forget the mind, and he who studies the mind cannot forget the nervous system. Much of the psychology of the past has proved to be really nothing more than the physiology of the organs of special sense; and the whole weight of the body of ideas which cybernetics is introducing into psychology concerns the physiology and anatomy of the highly specialized cortical areas connecting with these organs of special sense. From the beginning, we have anticipated that the problem of the perception of *Gestalt*, or of the perceptual formation of universals, would prove to be of this nature. What is the mechanism by which we recognize a square as a square, irrespective of its position, its size, and its orientation? To assist us in such matters and to inform them of whatever use might be made of our concepts for their assistance, we had among us such psychologists as Professor Klüver of the University of Chicago, the late Dr. Kurt Lewin of the Massachusetts Institute of Technology, and Dr. M. Ericsson of New York.

As to sociology and anthropology, it is manifest that the importance of information and communication as mechanisms of organization proceeds beyond the individual into the community. On the one hand, it is completely impossible to understand social communities such as those of ants without a thorough investigation of their means of communication, and we were fortunate enough to have the aid of Dr. Schneirla in this matter. For the similar problems of human organization, we sought help from the anthropologists Drs. Bateson and Margaret Mead; while Dr. Morgenstern of the Institute for Advanced Study was our adviser in the significant field of social organization belonging to economic theory. His very important joint book on games with Dr. von Neumann, by the way, represents a most interesting study of social organization from the point of view of methods closely related to, although distinct from, the subject matter of cybernetics. Dr. Lewin and others represented the newer work on the theory of opinion sampling and the practice of opinion making, and Dr. F. C. S. Northrup was interested in assaying the philosophical significance of our work.

This does not purport to be a complete list of our group. We also enlarged the group to contain more engineers and mathematicians such as Bigelow and Savage, more neuroanatomists and neurophysiologists such as von Bonin and Lloyd, and so on. Our first meeting, held in the spring of 1946, was largely devoted to didactic papers by those of us who had been present at the Princeton meeting and to a general assessment of the importance of the field by all present. It was the sense of the meeting that the ideas behind cybernetics were sufficiently important and interesting to those present to warrant a continuation of our meetings at intervals of six months; and that before the next full meeting, we should have a small meeting for the benefit of the less mathematically trained to explain to them in as simple language as possible the nature of the mathematical concepts involved.

In the summer of 1946, I returned to Mexico with the support of the Rockefeller Foundation and the hospitality of the Instituto Nacional de Cardiología to continue the collaboration between Dr. Rosenblueth and myself. This time we decided to take a nervous problem directly from the topic of feedback and to see what we could do with it experimentally. We chose the cat as our experimental animal, and the quadriceps extensor femoris as the muscle to study. We cut the attachment of the muscle, fixed it to a lever under known tension, and recorded its contractions isometrically or isotonically. We also used an oscillograph to record the simultaneous electrical changes in the muscle itself. We worked chiefly with cats, first decerebrated under ether anesthesia and later made spinal by a thoracic transection of the cord. In many cases, strychnine was used to increase the reflex responses. The muscle was loaded to the point where a tap would set it into a periodic pattern of contraction, which is called *clonus* in the language of the physiologist. We observed this pattern of contraction, paying attention to the physiological condition of the cat, the load on the muscle, the frequency of oscillation, the base level of the oscillation, and its amplitude. These we tried to analyze as we should analyze a mechanical or electrical system exhibiting the same pattern of hunting. We employed, for example, the methods of MacColl's

book on servomechanisms.   This is not the place to discuss the full
significance of our results, which we are now repeating and preparing
to write up for publication.   However, the following statements are
either established or very probable: that the frequency of clonic
oscillation is much less sensitive to changes of the loading conditions
than we had expected, and that it is much more nearly determined
by the constants of the closed arc (efferent-nerve)–muscle–(kines-
thetic-end-body)–(afferent-nerve)–(central-synapse)–(efferent-nerve)
than by anything else.   This circuit is not even approximately a
circuit of linear operators if we take as our base of linearity the number of impulses transmitted by the efferent nerve per second, but
seems to become much more nearly so if we replace the number of
impulses by its logarithm.   This corresponds to the fact that the
form of the envelope of stimulation of the efferent nerve is not nearly
sinusoidal, but that the logarithm of this curve is much more nearly
sinusoidal; while in a linear oscillating system with constant energy
level, the form of the curve of stimulation must be sinusoidal in all
except a set of cases of zero probability.   Again, the notions of
facilitation and inhibition are much more nearly multiplicative than
additive in nature.   For example, a complete inhibition means a
multiplication by zero, and a partial inhibition means a multiplication by a small quantity.   It is these notions of inhibition and
facilitation which have been used [1] in the discussion of the reflex arc.
Furthermore, the synapse is a coincidence-recorder, and the outgoing
fiber is stimulated only if the number of incoming impulses in a small
summation time exceeds a certain threshold.   If this threshold is
low enough in comparison with the full number of incoming synapses,
the synaptic mechanism serves to multiply probabilities, and that it
can be even an approximately linear link is possible only in a
logarithmic system.   This approximate logarithmicity of the synapse
mechanism is certainly allied to the approximate logarithmicity of
the Weber-Fechner law of sensation intensity, even though this law
is only a first approximation.

   The most striking point is that on this logarithmic basis, and with
data obtained from the conduction of single pulses through the
various elements of the neuromuscular arc, we were able to obtain
very fair approximations to the actual periods of clonic vibration,
using the technique already developed by the servo engineers for the
determination of the frequencies of hunting oscillations in feedback
systems which have broken down.   We obtained theoretical oscilla-

[1] Unpublished articles on clonus from the Instituto Nacional de Cardiología,
Mexico.

tions of about 13.9 per second, in cases where the observed oscillations varied between frequencies of 7 and 30, but generally remained within a range varying somewhere between 12 and 17. Under the circumstances, this agreement is excellent.

The frequency of clonus is not the only important phenomenon which we may observe: there is also a relatively slow change in basal tension, and an even slower change in amplitude. These phenomena are certainly by no means linear. However, sufficiently slow changes in the constants of a linear oscillating system may be treated to a first approximation as though they were infinitely slow, and as though over each part of the oscillation the system behaved as it would if its parameters were those belonging to it at the time. This is the method known in other branches of physics as that of secular perturbations. It may be used to study the problems of base level and amplitude of clonus. While this work has not yet been completed, it is clear that it is both possible and promising. There is a strong suggestion that though the timing of the main arc in clonus proves it to be a two-neuron arc, the amplification of impulses in this arc is variable in one and perhaps in more points, and that some part of this amplification may be affected by slow, multineuron processes which run much higher in the central nervous system than the spinal chain primarily responsible for the timing of clonus. This variable amplification may be affected by the general level of central activity, by the use of strychnine or of anesthetics, by decerebration, and by many other causes.

These were the main results presented by Dr. Rosenblueth and myself at the Macy meeting held in the autumn of 1946, and in a meeting of the New York Academy of Sciences held at the same time for the purpose of diffusing the notions of cybernetics over a larger public. While we were pleased with our results, and fully convinced of the general practicability of work in this direction, we felt nevertheless that the time of our collaboration had been too brief, and that our work had been done under too much pressure to make it desirable to publish without further experimental confirmation. This confirmation—which naturally might amount to a refutation—we are now seeking in the summer and autumn of 1947.

The Rockefeller Foundation had already given Dr. Rosenblueth a grant for the equipment of a new laboratory building at the Instituto Nacional de Cardiología. We felt that the time was now ripe for us to go jointly to them—that is, to Dr. Warren Weaver, in charge of the department of physical sciences, and to Dr. Robert Morison, in charge of the department of medical sciences—to establish the basis of a long-time scientific collaboration, in order to carry on our program at a more leisurely and healthy pace. In this we were enthusiastically backed by our respective institutions. Dr. George Harrison, Dean of Science, was the chief representative of the Massachusetts Institute of Technology during these negotiations, while Dr. Ignacio Chávez spoke for his institution, the Instituto Nacional de Cardiología. During the negotiations, it became clear that the laboratory center of the joint activity should be at the Instituto, both in order to avoid the duplication of laboratory equipment and to further the very real interest the Rockefeller Foundation has shown in the establishment of scientific centers in Latin America. The plan finally adopted was for five years, during which I should spend six months of every other year at the Instituto, while Dr. Rosenblueth would spend six months of the intervening years at the Institute. The time at the Instituto is to be devoted to the obtaining and elucidation of experimental data pertaining to cybernetics, while the intermediate years are to be devoted to more theoretical research and, above all, to the very difficult problem of devising, for people wishing to go into this new field, a scheme of training which will secure for them both the necessary mathematical, physical, and engineering background and the proper acquaintance with biological, psychological, and medical techniques.

In the spring of 1947, Dr. McCulloch and Mr. Pitts did a piece of work of considerable cybernetic importance. Dr. McCulloch had been given the problem of designing an apparatus to enable the blind to read the printed page by ear. The production of variable tones by type through the agency of a photocell is an old story, and can be effected by any number of methods; the difficult point is to make the pattern of the sound substantially the same when the pattern of the letters is given, whatever the size. This is a definite analogue of the problem of the perception of form, of *Gestalt*, which allows us to recognize a square as a square through a large number of changes of size and of orientation. Dr. McCulloch's device involved a selective reading of the type imprint for a set of different magnifications. Such a selective reading can be performed automatically as a scanning process. This scanning, to allow a comparison between a figure and a given standard figure of fixed but different size, was a device which I had already suggested at one of the Macy meetings. A diagram of the apparatus by which the selective reading was done came to the attention of Dr. von Bonin, who immediately asked, "Is this a diagram of the fourth layer of the visual cortex of the brain?" Acting on this suggestion, Dr. McCulloch, with the assistance of

Mr. Pitts, produced a theory tying up the anatomy and the physiology of the visual cortex, and in this theory the operation of scanning over a set of transformations plays an important part. This was presented in the spring of 1947, both at the Macy meeting and at a meeting of the New York Academy of Sciences. Finally, this scanning process involves a certain periodic time, which corresponds to what we call the "time of sweep" in ordinary television. There are various anatomic clues to this time in the length of the chain of consecutive synapses necessary to run around one cycle of performance. These yield a time of the order of a tenth of a second for a complete performance of the cycle of operations, and this is the approximate period of the so-called "alpha rhythm" of the brain. Finally, the alpha rhythm, on quite other evidence, has already been conjectured to be of visual origin and to be important in the process of form perception.

In the spring of 1947, I received an invitation to participate in a mathematical conference in Nancy on problems arising from harmonic analysis. I accepted and, on my voyage there and back, spent a total of three weeks in England, chiefly as a guest of my old friend Professor J. B. S. Haldane. I had an excellent chance to meet most of those doing work on ultra-rapid computing machines, especially at Manchester and at the National Physical Laboratories at Teddington, and above all to talk over the fundamental ideas of cybernetics with Mr. Turing at Teddington. I also visited the Psychological Laboratory at Cambridge, and had a very good chance to discuss the work that Professor F. C. Bartlett and his staff were doing on the human element in control processes involving such an element. I found the interest in cybernetics about as great and well informed in England as in the United States, and the engineering work excellent, though of course limited by the smaller funds available. I found much interest and understanding of its possibility in many quarters, and Professors Haldane, H. Levy, and Bernal certainly regarded it as one of the most urgent problems on the agenda of science and scientific philosophy. I did not find, however, that as much progress had been made in unifying the subject and in pulling the various threads of research together as we had made at home in the States.

In France, the meeting at Nancy on harmonic analysis contained a number of papers uniting statistical ideas and ideas from communication engineering in a manner wholly in conformity with the point of view of cybernetics. Here I must mention especially the names of M. Blanc-Lapierre and M. Loève. I found also a considerable interest in the subject on the part of mathematicians, physiologists, and physical chemists, particularly with regard to its thermodynamic aspects in so far as they touch the more general problem of the nature of life itself. Indeed, I had discussed that subject in Boston, before my departure, with Professor Szent-Györgyi, the Hungarian biochemist, and had found his ideas concordant with my own.

One event during my French visit is particularly worth while noting here. My colleague, Professor G. de Santillana of M.I.T., introduced me to M. Freymann, of the firm of Hermann et Cie, and he requested of me the present book. I am particularly glad to receive his invitation, as M. Freymann is a Mexican, and the writing of the present book, as well as a good deal of the research leading up to it, has been done in Mexico.

As I have already hinted, one of the directions of work which the realm of ideas of the Macy meetings has suggested concerns the importance of the notion and the technique of communication in the social system. It is certainly true that the social system is an organization like the individual, that it is bound together by a system of communication, and that it has a dynamics in which circular processes of a feedback nature play an important part. This is true, both in the general fields of anthropology and of sociology and in the more specific field of economics; and the very important work, which we have already mentioned, of von Neumann and Morgenstern on the theory of games enters into this range of ideas. On this basis, Drs. Gregory Bateson and Margaret Mead have urged me, in view of the intensely pressing nature of the sociological and economic problems of the present age of confusion, to devote a large part of my energies to the discussion of this side of cybernetics.

Much as I sympathize with their sense of the urgency of the situation, and much as I hope that they and other competent workers will take up problems of this sort, which I shall discuss in a later chapter of this book, I can share neither their feeling that this field has the first claim on my attention, nor their hopefulness that sufficient progress can be registered in this direction to have an appreciable therapeutic effect in the present diseases of society. To begin with, the main quantities affecting society are not only statistical, but the runs of statistics on which they are based are excessively short. There is no great use in lumping under one head the economics of steel industry before and after the introduction of the Bessemer process, nor in comparing the statistics of rubber production before and after the burgeoning of the automobile industry and the cultivation of *Hevea* in Malaya.   Neither is there any important point in running statistics of the incidence of venereal disease in a single table which covers both the period before and that after the introduction of salvarsan, unless for the specific purpose of studying the effectiveness of this drug.   For a good statistic of society, we need long runs *under essentially constant conditions,* just as for a good resolution of light we need a lens with a large aperture.   The effective aperture of a lens is not appreciably increased by augmenting its nominal aperture, *unless the lens is made of a material so homogeneous that the delay of light in different parts of the lens conforms to the proper designed amount by less than a small part of a wavelength.   Similarly, the advantage of long runs of statistics under widely varying conditions is specious and spurious.*   Thus the human sciences are very poor testing-grounds for a new mathematical technique: as poor as the statistical mechanics of a gas would be to a being of the order of size of a molecule, to whom the fluctuations which we ignore from a larger standpoint would be precisely the matters of greatest interest. Moreover, in the absence of reasonably safe routine numerical techniques, the element of the judgment of the expert in determining the estimates to be made of sociological, anthropological, and economic quantities is so great that it is no field for a newcomer who has not yet had the bulk of experience which goes to make up the expert.   I may remark parenthetically that the modern apparatus of the theory of small samples, once it goes beyond the determination of its own specially defined parameters and becomes a method for positive statistical inference in new cases, does not inspire me with any confidence unless it is applied by a statistician by whom the main elements of the dynamics of the situation are either explicitly known or implicitly felt.

I have just spoken of a field in which my expectations of cybernetics are definitely tempered by an understanding of the limitations of the data which we may hope to obtain.   There are two other fields where I ultimately hope to accomplish something practical with the aid of cybernetic ideas, but in which this hope must wait on further developments.   One of these is the matter of prostheses for lost or paralyzed limbs.   As we have seen in the discussion of *Gestalt,* the ideas of communication engineering have already been applied by McCulloch to the problem of the replacement of lost senses, in the construction of an instrument to enable the blind to read print by hearing.   Here the instrument suggested by McCulloch takes over quite explicitly some of the functions not only of the eye but of the visual cortex.   There is a manifest possibility of doing something similar in the case of artificial limbs.   The loss of a segment of limb implies not only the loss of the purely passive support of the missing segment or its value as mechanical extension of the stump, and the loss of the contractile power of its muscles, but implies as well the loss of all cutaneous and kinesthetic sensations originating in it. The first two losses are what the artificial-limbmaker now tries to replace.   The third has so far been beyond his scope.   In the case of a simple peg leg, this is not important: the rod that replaces the missing limb has no degrees of freedom of its own, and the kinesthetic mechanism of the stump is fully adequate to report its own position and velocity.   This is not the case with the articulated limb with a mobile knee and ankle, thrown ahead by the patient with the aid of his remaining musculature.   He has no adequate report of their position and motion, and this interferes with his sureness of step on an irregular terrain.   There does not seem to be any insuperable difficulty in equipping the artificial joints and the sole of the artificial foot with strain or pressure gauges, which are to register electrically or otherwise, say through vibrators, on intact areas of skin.   The present artificial limb removes some of the paralysis caused by the amputation but leaves the ataxis.   With the use of proper receptors, much of this ataxia should disappear as well, and the patient should be able to learn reflexes, such as those we all use in driving a car, which should enable him to step out with a much surer gait.   What we have said about the leg should apply with even more force to the arm, where the figure of the manikin familiar to all readers of books of neurology shows that the sensory loss in an amputation of the thumb alone is considerably greater than the sensory loss even in a hip-joint amputation.

I have made an attempt to report these considerations to the proper authorities, but up to now I have not been able to accomplish much.   I do not know whether the same ideas have already emanated from other sources, nor whether they have been tried out and found technically impracticable.   In case they have not yet received a thorough practical consideration, they should receive one in the immediate future.

Let me now come to another point which I believe to merit attention.   It has long been clear to me that the modern ultra-rapid computing machine was in principle an ideal central nervous system to an apparatus for automatic control; and that its input and output need not be in the form of numbers or diagrams but might very well be, respectively, the readings of artificial sense organs, such as photoelectric cells or thermometers, and the performance of motors or solenoids. With the aid of strain gauges or similar agencies to read the performance of these motor organs and to report, to "feed back," to the central control system as an artificial kinesthetic sense, we are already in a position to construct artificial machines of almost any degree of elaborateness of performance. Long before Nagasaki and the public awareness of the atomic bomb, it had occurred to me that we were here in the presence of another social potentiality of unheard-of importance for good and for evil. The automatic factory and the assembly line without human agents are only so far ahead of us as is limited by our willingness to put such a degree of effort into their engineering as was spent, for example, in the development of the technique of radar in the Second World War.[1]

I have said that this new development has unbounded possibilities for good and for evil. For one thing, it makes the metaphorical dominance of the machines, as imagined by Samuel Butler, a most immediate and non-metaphorical problem. It gives the human race a new and most effective collection of mechanical slaves to perform its labor. Such mechanical labor has most of the economic properties of slave labor, although, unlike slave labor, it does not involve the direct demoralizing effects of human cruelty. However, any labor that accepts the conditions of competition with slave labor accepts the conditions of slave labor, and is essentially slave labor. The key word of this statement is *competition*. It may very well be a good thing for humanity to have the machine remove from it the need of menial and disagreeable tasks, or it may not. I do not know. It cannot be good for these new potentialities to be assessed in the terms of the market, of the money they save; and it is precisely the terms of the open market, the "fifth freedom," that have become the shibboleth of the sector of American opinion represented by the National Association of Manufacturers and the Saturday Evening Post. I say American opinion, for as an American, I know it best, but the hucksters recognize no national boundary.

Perhaps I may clarify the historical background of the present situation if I say that the first industrial revolution, the revolution of the "dark satanic mills," was the devaluation of the human arm by the competition of machinery. There is no rate of pay at which a United States pick-and-shovel laborer can live which is low enough to compete with the work of a steam shovel as an excavator. The modern industrial revolution is similarly bound to devalue the human brain, at least in its simpler and more routine decisions. Of course, just as the skilled carpenter, the skilled mechanic, the skilled

[1] *Fortune*, **32**, 139–147 (October); 163–169 (November, 1945).

dressmaker have in some degree survived the first industrial revolution, so the skilled scientist and the skilled administrator may survive the second.   However, taking the second revolution as accomplished, the average human being of mediocre attainments or less has nothing to sell that it is worth anyone's money to buy.

The answer, of course, is to have a society based on human values other than buying or selling.   To arrive at this society, we need a good deal of planning and a good deal of struggle, which, if the best comes to the best, may be on the plane of ideas, and otherwise—who knows?   I thus felt it my duty to pass on my information and understanding of the position to those who have an active interest in the conditions and the future of labor, that is, to the labor unions.   I did manage to make contact with one or two persons high up in the C.I.O., and from them I received a very intelligent and sympathetic hearing.   Further than these individuals, neither I nor any of them was able to go.   It was their opinion, as it had been my previous observation and information, both in the United States and in England, that the labor unions and the labor movement are in the hands of a highly limited personnel, thoroughly well trained in the specialized problems of shop stewardship and disputes concerning wages and conditions of work, and totally unprepared to enter into the larger political, technical, sociological, and economic questions which concern the very existence of labor.   The reasons for this are easy enough to see: the labor union official generally comes from the exacting life of a workman into the exacting life of an administrator without any opportunity for a broader training; and for those who have this training, a union career is not generally inviting; nor, quite naturally, are the unions receptive to such people.

Those of us who have contributed to the new science of cybernetics thus stand in a moral position which is, to say the least, not very comfortable.   We have contributed to the initiation of a new science which, as I have said, embraces technical developments with great possibilities for good and for evil.   We can only hand it over into the world that exists about us, and this is the world of Belsen and Hiroshima.   We do not even have the choice of suppressing these new technical developments.   They belong to the age, and the most any of us can do by suppression is to put the development of the subject into the hands of the most irresponsible and most venal of our engineers.   The best we can do is to see that a large public understands the trend and the bearing of the present work, and to confine our personal efforts to those fields, such as physiology and psychology, most remote from war and exploitation.   As we have seen, there are those who hope that the good of a better understanding of man and society which is offered by this new field of work may anticipate and outweigh the incidental contribution we are making to the concentration of power (which is always concentrated, by its very conditions of existence, in the hands of the most unscrupulous).   I write in 1947, and I am compelled to say that it is a very slight hope.

The author wishes to express his gratitude to Mr. Walter Pitts, Mr. Oliver Selfridge, Mr. Georges Dubé, and Mr. Frederic Webster for aid in correcting the manuscript and preparing the material for publication.

Instituto Nacional de Cardiología,
Ciudad de México

November, 1947

I

Newtonian and Bergsonian Time

There is a little hymn or song familiar to every German child.
It goes:

> „ Weisst du, wieviel Sternlein stehen
> An dem blauen Himmelszelt?
> Weisst du, wieviel Wolken gehen
> Weithin über alle Welt?
> Gott, der Herr, hat sie gezählet
> Dass ihm auch nicht eines fehlet
> An der ganzen, grossen Zahl."
>
> W. Hey

In English this says: "Knowest thou how many stars stand in the
blue tent of heaven? Knowest thou how many clouds pass far over
the whole world? The Lord God hath counted them, that not one of
the whole great number be lacking."

This little song is an interesting theme for the philosopher and the
historian of science, in that it puts side by side two sciences which
have the one similarity of dealing with the heavens above us, but
which in almost every other respect offer an extreme contrast.
Astronomy is the oldest of the sciences, while meteorology is among
the youngest to begin to deserve the name. The more familiar
astronomical phenomena can be predicted for many centuries, while
a precise prediction of tomorrow's weather is generally not easy and
in many places very crude indeed.

To go back to the poem, the answer to the first question is that,
within limits, we do know how many stars there are. In the first
place, apart from minor uncertainties concerning some of the double

and variable stars, a star is a definite object, eminently suitable for counting and cataloguing; and if a human *Durchmusterung* of the stars—as we call these catalogues—stops short for stars less intense than a certain magnitude, there is nothing too repugnant to us in the idea of a divine *Durchmusterung* going much further.

On the other hand, if you were to ask the meteorologist to give you a similar *Durchmusterung* of the clouds, he might laugh in your face, or he might patiently explain that in all the language of meteorology there is no such thing as a cloud, defined as an object with a quasi-permanent identity; and that if there were, he neither possesses the facilities to count them, nor is he in fact interested in counting them. A topologically inclined meteorologist might perhaps define a cloud as a connected region of space in which the density of the part of the water content in the solid or liquid state exceeds a certain amount, but this definition would not be of the slightest value to anyone, and would at most represent an extremely transitory state. What really concerns the meteorologist is some such statistical statement as, "Boston: January 17, 1950: Sky 38% overcast: Cirrocumulus."

There is of course a branch of astronomy which deals with what may be called cosmic meteorology: the study of galaxies and nebulae and star clusters and their statistics, as pursued for example by Chandrasekhar, but this is a very young branch of astronomy, younger than meteorology itself, and is something outside the tradition of classical astronomy. This tradition, apart from its purely classificatory, *Durchmusterung* aspects, was originally concerned rather with the solar system than with the world of the fixed stars. It is the astronomy of the solar system which is that chiefly associated with the names of Copernicus, Kepler, Galileo, and Newton, and which was the wet nurse of modern physics.

It is indeed an ideally simple science. Even before the existence of any adequate dynamical theory, even as far back as the Babylonians, it was realized that eclipses occurred in regular predictable cycles, extending backward and forward over time. It was realized that time itself could better be measured by the motion of the stars in their courses than in any other way. The pattern for all events in the solar system was the revolution of a wheel or a series of wheels, whether in the form of the Ptolemaic theory of epicycles or the Copernican theory of orbits, and in any such theory the future after a fashion repeats the past. The music of the spheres is a palindrome, and the book of astronomy reads the same backward as forward. There is no difference save of initial positions and directions between the motion of an orrery turned forward and one run in reverse. Finally, when all this was reduced by Newton to a formal set of postulates and a closed mechanics, the fundamental laws of this mechanics were unaltered by the transformation of the time variable t into its negative.

Thus if we were to take a motion picture of the planets, speeded up to show a perceptible picture of activity, and were to run the film backward, it would still be a possible picture of planets conforming to the Newtonian mechanics. On the other hand, if we were to take a motion-picture photograph of the turbulence of the clouds in a thunderhead and reverse it, it would look altogether wrong. We should see downdrafts where we expect updrafts, turbulence growing coarser in texture, lightning preceding instead of following the changes of cloud which usually precede it, and so on indefinitely.

What is the difference between the astronomical and the meteorological situation which brings about all these differences, and in particular the difference between the apparent reversibility of astronomical time and the apparent irreversibility of meteorological time? In the first place, the meteorological system is one involving a vast number of approximately equal particles, some of them very closely coupled to one another, while the astronomical system of the solar universe contains only a relatively small number of particles, greatly diverse in size and coupled with one another in a sufficiently loose way that the second-order coupling effects do not change the general aspect of the picture we observe, and the very high order coupling effects are completely negligible. The planets move under conditions more favorable to the isolation of a certain limited set of forces than those of any physical experiment we can set up in the laboratory. Compared with the distances between them, the planets, and even the sun, are very nearly points. Compared with the elastic and plastic deformations they suffer, the planets are either very nearly rigid bodies, or, where they are not, their internal forces are at any rate of a relatively slight significance where the relative motion of their centers is concerned. The space in which they move is almost perfectly free from impeding matter; and in their mutual attraction, their masses may be considered to lie very nearly at their centers and to be constant. The departure of the law of gravity from the inverse square law is most minute. The positions, velocities, and masses of the bodies of the solar system are extremely well known at any time, and the computation of their future and past positions, while not easy in detail, is easy and precise in principle. On the other hand, in meteorology, the number of particles concerned is so enormous that an accurate record of their initial positions and velocities is utterly impossible; and if this record were actually made and their future positions and velocities computed, we should have nothing but an impenetrable mass of figures which would need a radical reinterpretation before it could be of any service to us.   The terms "cloud," "temperature," "turbulence," etc., are all terms referring not to one single physical situation but to a distribution of possible situations of which only one actual case is realized.   If all the readings of all the meteorological stations on earth were simultaneously taken, they would not give a billionth part of the data necessary to characterize the actual state of the atmosphere from a Newtonian point of view.   They would only give certain constants consistent with an infinity of different atmospheres, and at most, together with certain a *priori* assumptions, capable of giving, as a probability distribution, a measure over the set of possible atmospheres.   Using the Newtonian laws, or any other system of causal laws whatever, all that we can predict at any future time is a probability distribution of the constants of the system, and even this predictability fades out with the increase of time.

Now, even in a Newtonian system, in which time is perfectly reversible, questions of probability and prediction lead to answers asymmetrical as between past and future, because the questions to which they are answers are asymmetrical.   If I set up a physical experiment, I bring the system I am considering from the past into the present in such a way that I fix certain quantities and have a reasonable right to assume that certain other quantities have known statistical distributions.   I then observe the statistical distribution of results after a given time.   This is not a process which I can reverse. In order to do so, it would be necessary to pick out a fair distribution of systems which, without intervention on our part, would end up within certain statistical limits, and find out what the antecedent conditions were a given time ago.   However, for a system starting from an unknown position to end up in any tightly defined statistical range is so rare an occurrence that we may regard it as a miracle, and we cannot base our experimental technique on awaiting and counting miracles.   In short, we are directed in time, and our relation to the future is different from our relation to the past.   All our questions are conditioned by this asymmetry, and all our answers to these questions are equally conditioned by it.

A very interesting astronomical question concerning the direction of time comes up in connection with the time of astrophysics, in which we are observing remote heavenly bodies in a single observation, and in which there seems to be no unidirectionalness in the nature of our experiment.   Why then does the unidirectional thermodynamics which is based on experimental terrestrial observations stand us in such good stead in astrophysics?   The answer is interesting and not too obvious.   Our observations of the stars are through the agency of light, of rays or particles emerging from the observed object and perceived by us.   We can perceive incoming light, but can not perceive outgoing light, or at least the perception of outgoing light is not achieved by an experiment as simple and direct as that of incoming light.   In the perception of incoming light, we end up with the eye or a photographic plate.   We condition these for the reception of images by putting them in a state of insulation for some time past: we dark-condition the eye to avoid after-images, and we wrap our plates in black paper to prevent halation.   It is clear that only such an eye and only such plates are any use to us: if we were given to pre-images, we might as well be blind; and if we had to put our plates in black paper after we use them and develop them before using, photography would be a very difficult art indeed.   This being the case, we can see those stars radiating to us and to the whole world; while if there are any stars whose evolution is in the reverse direction, they will attract radiation from the whole heavens, and even this attraction from us will not be perceptible in any way, in view of the fact that we already know our own past but not our future.   Thus the part of the universe which we see must have its past-future relations, as far as the emission of radiation is concerned, concordant with our own.   The very fact that we see a star means that its thermodynamics is like our own.

Indeed, it is a very interesting intellectual experiment to make the fantasy of an intelligent being whose time should run the other way to our own.   To such a being, all communication with us would be impossible.   Any signal he might send would reach us with a logical stream of consequents from his point of view, antecedents from ours.   These antecedents would already be in our experience, and would have served to us as the natural explanation of his signal, without presupposing an intelligent being to have sent it.   If he drew us a square, we should see the remains of his figure as its precursors, and it would seem to be the curious crystallization—always perfectly explainable—of these remains.   Its meaning would seem to be as fortuitous as the faces we read into mountains and cliffs.   The drawing of the square would appear to us as a catastrophe—sudden indeed, but explainable by natural laws—by which that square would cease to exist.   Our counterpart would have exactly similar ideas concerning us. *Within any world with which we can communicate, the direction of time is uniform.*

To return to the contrast between Newtonian astronomy and meteorology: most sciences lie in an intermediate position, but most are rather nearer to meteorology than to astronomy. Even astronomy, as we have seen, contains a cosmic meteorology. It contains as well that extremely interesting field studied by Sir George Darwin, and known as the theory of tidal evolution. We have said that we can treat the relative movements of the sun and the planets as the movements of rigid bodies, but this is not quite the case. The earth, for example, is nearly surrounded by oceans. The water nearer the moon than the center of the earth is more strongly attracted to the moon than the solid part of the earth, and the water on the other side is less strongly attracted. This relatively slight effect pulls the water into two hills, one under the moon and one opposite to the moon. In a perfectly liquid sphere, these hills could follow the moon around the earth with no great dispersal of energy, and consequently would remain almost precisely under the moon and opposite to the moon. They would consequently have a pull on the moon which would not greatly influence the angular position of the moon in the heavens. However, the tidal wave they produce on the earth gets tangled up and delayed on coasts and in shallow seas such as the Bering Sea and the Irish Sea. It consequently lags behind the position of the moon, and the forces producing this are largely turbulent, dissipative forces, of a character much like the forces met in meteorology, and need a statistical treatment. Indeed, oceanography may be called the meteorology of the hydrosphere rather than of the atmosphere.

These frictional forces drag the moon back in its course about the earth and accelerate the rotation of the earth forward. They tend to bring the lengths of the month and of the day ever closer to one another. Indeed, the day of the moon is the month, and the moon always presents nearly the same face to the earth. It has been suggested that this is the result of an ancient tidal evolution, when the moon contained some liquid or gas or plastic material which could give under the earth's attraction, and in so giving could dissipate large amounts of energy. This phenomenon of tidal evolution is not confined to the earth and the moon but may be observed to some degree throughout all gravitating systems. In ages past it has seriously modified the face of the solar system, though in anything like historic times this modification is slight compared with the "rigid-body" motion of the planets of the solar system.

Thus even gravitational astronomy involves frictional processes that run down. There is not a single science which conforms precisely to the strict Newtonian pattern. The biological sciences certainly have their full share of one-way phenomena. Birth is not the exact reverse of death, nor is anabolism—the building up of tissues—the exact reverse of catabolism—their breaking down. The division of cells does not follow a pattern symmetrical in time, nor does the union of the germ cells to form the fertilized ovum. The individual is an arrow pointed through time in one way, and the race is equally directed from the past into the future.

The record of paleontology indicates a definite long-time trend, interrupted and complicated though it might be, from the simple to the complex. By the middle of the last century this trend had become apparent to all scientists with an honestly open mind, and it is no accident that the problem of discovering its mechanisms was carried ahead through the same great step by two men working at about the same time: Charles Darwin and Alfred Wallace. This step was the realization that a mere fortuitous variation of the individuals of a species might be carved into the form of a more or less one-directional or few-directional progress for each line by the varying degrees of viability of the several variations, either from the point of view of the individual or of the race. A mutant dog without legs will certainly starve, while a long thin lizard that has developed the mechanism of crawling on its ribs may have a better chance for survival if it has clean lines and is freed from the impeding projections of limbs. An aquatic animal, whether fish, lizard, or mammal, will swim better with a fusiform shape, powerful body muscles, and a posterior appendage which will catch the water; and if it is dependent for its food on the pursuit of swift prey, its chances of survival may depend on its assuming this form.

Darwinian evolution is thus a mechanism by which a more or less fortuitous variability is combined into a rather definite pattern. Darwin's principle still holds today, though we have a much better knowledge of the mechanism on which it depends. The work of Mendel has given us a far more precise and discontinuous view of heredity than that held by Darwin, while the notion of mutation, from the time of de Vries on, has completely altered our conception of the statistical basis of mutation. We have studied the fine anatomy of the chromosome and have localized the gene on it. The list of modern geneticists is long and distinguished. Several of these, such as Haldane, have made the statistical study of Mendelianism an effective tool for the study of evolution.

We have already spoken of the tidal evolution of Sir George Darwin, Charles Darwin's son. Neither the connection of the idea of the son with that of the father nor the choice of the name "evolution" is fortuitous. In tidal evolution as well as in the origin of species, we have a mechanism by means of which a fortuitous variability, that of the random motions of the waves in a tidal sea and of the molecules of the water, is converted by a dynamical process into a pattern of development which reads in one direction. The theory of tidal evolution is quite definitely an astronomical application of the elder Darwin.

The third of the dynasty of Darwins, Sir Charles, is one of the authorities on modern quantum mechanics. This fact may be fortuitous, but it nevertheless represents an even further invasion of Newtonian ideas by ideas of statistics. The succession of names Maxwell-Boltzmann-Gibbs represents a progressive reduction of thermodynamics to statistical mechanics: that is, a reduction of the phenomena concerning heat and temperature to phenomena in which a Newtonian mechanics is applied to a situation in which we deal not with a single dynamical system but with a statistical distribution of dynamical systems; and in which our conclusions concern not all such systems but an overwhelming majority of them. About the year 1900, it became apparent that there was something seriously wrong with thermodynamics, particularly where it concerned radiation. The ether showed much less power to absorb radiations of high frequency—as shown by the law of Planck—than any existing mechanization of radiation theory had allowed. Planck gave a quasi-atomic theory of radiation—the quantum theory—which accounted satisfactorily enough for these phenomena, but which was at odds with the whole remainder of physics; and Niels Bohr followed this up with a similarly *ad hoc* theory of the atom. Thus Newton and Planck-Bohr formed, respectively, the thesis and antithesis of a Hegelian antinomy. The synthesis is the statistical theory discovered by Heisenberg in 1925, in which the statistical Newtonian dynamics of Gibbs is replaced by a statistical theory very similar to that of Newton and Gibbs for large-scale phenomena, but in which the complete collection of data for the present and the past is not sufficient to predict the future more than statistically. It is thus not too much to say that not only the Newtonian astronomy but even the Newtonian physics has become a picture of the average results of a statistical situation, and hence an account of an evolutionary process.

This transition from a Newtonian, reversible time to a Gibbsian,

irreversible time has had its philosophical echoes. Bergson emphasized the difference between the reversible time of physics, in which
nothing new happens, and the irreversible time of evolution and
biology, in which there is always something new. The realization
that the Newtonian physics was not the proper frame for biology
was perhaps the central point in the old controversy between vitalism
and mechanism; although this was complicated by the desire to
conserve in some form or other at least the shadows of the soul and
of God against the inroads of materialism. In the end, as we have
seen, the vitalist proved too much. Instead of building a wall
between the claims of life and those of physics, the wall has been
erected to surround so wide a compass that both matter and life
find themselves inside it. It is true that the matter of the newer
physics is not the matter of Newton, but it is something quite as
remote from the anthropomorphizing desires of the vitalists. The
chance of the quantum theoretician is not the ethical freedom of the
Augustinian, and Tyche is as relentless a mistress as Ananke.

The thought of every age is reflected in its technique. The civil
engineers of ancient days were land surveyors, astronomers, and
navigators; those of the seventeenth and early eighteenth centuries
were clockmakers and grinders of lenses. As in ancient times, the
craftsmen made their tools in the image of the heavens. A watch is
nothing but a pocket orrery, moving by necessity as do the celestial
spheres; and if friction and the dissipation of energy play a role in
it, they are effects to be overcome, so that the resulting motion of
the hands may be as periodic and regular as possible. The chief
technical result of this engineering after the model of Huyghens and
Newton was the age of navigation, in which for the first time it was
possible to compute longitudes with a respectable precision, and to
convert the commerce of the great oceans from a thing of chance and
adventure to a regular understood business. It is the engineering
of the mercantilists.

To the merchant succeeded the manufacturer, and to the chronom-
eter, the steam engine. From the Newcomen engine almost to the
present time, the central field of engineering has been the study of
prime movers. Heat has been converted into usable energy of
rotation and translation, and the physics of Newton has been
supplemented by that of Rumford, Carnot, and Joule. Thermodynamics makes its appearance, a science in which time is eminently
irreversible; and although the earlier stages of this science seem to
represent a region of thought almost without contact with the Newtonian dynamics, the theory of the conservation of energy and the

later statistical explanation of the Carnot principle or second law of thermodynamics or principle of the degradation of energy—that principle which makes the maximum efficiency obtainable by a steam engine depend on the working temperatures of the boiler and the condenser—all these have fused thermodynamics and the Newtonian dynamics into the statistical and the non-statistical aspects of the same science.

If the seventeenth and early eighteenth centuries are the age of clocks, and the later eighteenth and the nineteenth centuries constitute the age of steam engines, the present time is the age of communication and control. There is in electrical engineering a split which is known in Germany as the split between the technique of strong currents and the technique of weak currents, and which we know as the distinction between power and communication engineering. It is this split which separates the age just past from that in which we are now living. Actually, communication engineering can deal with currents of any size whatever and with the movement of engines powerful enough to swing massive gun turrets; what distinguishes it from power engineering is that its main interest is not economy of energy but the accurate reproduction of a signal. This signal may be the tap of a key, to be reproduced as the tap of a telegraph receiver at the other end; or it may be a sound transmitted and received through the apparatus of a telephone; or it may be the turn of a ship's wheel, received as the angular position of the rudder. Thus communication engineering began with Gauss, Wheatstone, and the first telegraphers. It received its first reasonably scientific treatment at the hands of Lord Kelvin, after the failure of the first transatlantic cable in the middle of the last century; and from the eighties on, it was perhaps Heaviside who did the most to bring it into a modern shape. The discovery of radar and its use in the Second World War, together with the exigencies of the control of anti-aircraft fire, have brought to the field a large number of well-trained mathematicians and physicists. The wonders of the automatic computing machine belong to the same realm of ideas, which was certainly never so actively pursued in the past as it is at the present day.

At every stage of technique since Daedalus or Hero of Alexandria, the ability of the artificer to produce a working simulacrum of a living organism has always intrigued people. This desire to produce and to study automata has always been expressed in terms of the living technique of the age. In the days of magic, we have the bizarre and sinister concept of the Golem, that figure of clay into which the Rabbi of Prague breathed life with the blasphemy of the Ineffable Name of God.   In the time of Newton, the automaton becomes the clockwork music box, with the little effigies pirouetting stiffly on top.   In the nineteenth century, the automaton is a glorified heat engine, burning some combustible fuel instead of the glycogen of the human muscles.   Finally, the present automaton opens doors by means of photocells, or points guns to the place at which a radar beam picks up an airplane, or computes the solution of a differential equation.

Neither the Greek nor the magical automaton lies along the main lines of the direction of development of the modern machine, nor do they seem to have had much of an influence on serious philosophic thought.   It is far different with the clockwork automaton.   This idea has played a very genuine and important role in the early history of modern philosophy, although we are rather prone to ignore it.

To begin with, Descartes considers the lower animals as automata. This is done to avoid questioning the orthodox Christian attitude that animals have no souls to be saved or damned.   Just how these living automata function is something that Descartes, so far as I know, never discusses.   However, the important allied question of the mode of coupling of the human soul, both in sensation and in will, with its material environment is one which Descartes does discuss, although in a very unsatisfactory manner.   He places this coupling in the one median part of the brain known to him, the pineal gland. As to the nature of his coupling—whether or not it represents a direct action of mind on matter and of matter on mind—he is none too clear.   He probably does regard it as a direct action in both ways, but he attributes the validity of human experience in its action on the outside world to the goodness and honesty of God.

The role attributed to God in this matter is unstable.   Either God is entirely passive, in which case it is hard to see how Descartes' explanation really explains anything, or He is an active participant, in which case it is hard to see how the guarantee given by His honesty can be anything but an active participation in the act of sensation.   Thus the causal chain of material phenomena is paralleled by a causal chain starting with the act of God, by which He produces in us the experiences corresponding to a given material situation. Once this is assumed, it is entirely natural to attribute the correspondence between our will and the effects it seems to produce in the external world to a similar divine intervention.   This is the path followed by the Occasionalists, Geulincx and Malebranche.   In

Spinoza, who is in many ways the continuator of this school, the doctrine of Occasionalism assumes the more reasonable form of asserting that the correspondence between mind and matter is that of two self-contained attributes of God; but Spinoza is not dynamically minded, and gives little or no attention to the mechanism of this correspondence.

This is the situation from which Leibniz starts, but Leibniz is as dynamically minded as Spinoza is geometrically minded. First, he replaces the pair of corresponding elements, mind and matter, by a continuum of corresponding elements: the monads. While these are conceived after the pattern of the soul, they include many instances which do not rise to the degree of self-consciousness of full souls, and which form part of that world which Descartes would have attributed to matter. Each of them lives in its own closed universe, with a perfect causal chain from the creation or from minus infinity in time to the indefinitely remote future; but closed though they are, they correspond one to the other through the pre-established harmony of God. Leibniz compares them to clocks which have so been wound up as to keep time together from the creation for all eternity. Unlike humanly made clocks, they do not drift into asynchronism; but this is due to the miraculously perfect workmanship of the Creator.

Thus Leibniz considers a world of automata, which, as is natural in a disciple of Huyghens, he constructs after the model of clockwork. Though the monads reflect one another, the reflection does not consist in a transfer of the causal chain from one to another. They are actually as self-contained as, or rather more self-contained than, the passively dancing figures on top of a music box. They have no real influence on the outside world, nor are they effectively influenced by it. As he says, they have no windows. The apparent organization of the world we see is something between a figment and a miracle. The monad is a Newtonian solar system writ small.

In the nineteenth century, the automata which are humanly constructed and those other natural automata, the animals and plants of the materialist, are studied from a very different aspect. The conservation and the degradation of energy are the ruling principles of the day. The living organism is above all a heat engine, burning glucose or glycogen or starch, fats, and proteins into carbon dioxide, water, and urea. It is the metabolic balance which is the center of attention; and if the low working temperatures of animal muscle attract attention as opposed to the high working temperatures of a heat engine of similar efficiency, this fact is pushed

into a corner and glibly explained by a contrast between the chemical
energy of the living organism and the thermal energy of the heat
engine.   All the fundamental notions are those associated with
energy, and the chief of these is that of potential.   The engineering
of the body is a branch of power engineering.   Even today, this is the
predominating point of view of the more classically minded, conservative physiologists; and the whole trend of thought of such
biophysicists as Rashevsky and his school bears witness to its
continued potency.

Today we are coming to realize that the body is very far from a
conservative system, and that its component parts work in an
environment where the available power is much less limited than
we have taken it to be.   The electronic tube has shown us that a
system with an outside source of energy, almost all of which is
wasted, may be a very effective agency for performing desired
operations, especially if it is worked at a low energy level.   We are
beginning to see that such important elements as the neurons, the
atoms of the nervous complex of our body, do their work under much
the same conditions as vacuum tubes, with their relatively small
power supplied from outside by the circulation, and that the book-
keeping which is most essential to describe their function is not one
of energy.   In short, the newer study of automata, whether in the
metal or in the flesh, is a branch of communication engineering, and
its cardinal notions are those of message, amount of disturbance or
"noise"—a term taken over from the telephone engineer—quantity
of information, coding technique, and so on.

In such a theory, we deal with automata effectively coupled to the
external world, not merely by their energy flow, their metabolism,
but also by a flow of impressions, of incoming messages, and of the
actions of outgoing messages.   The organs by which impressions are
received are the equivalents of the human and animal sense organs.
They comprise photoelectric cells and other receptors for light;
radar systems, receiving their own short Hertzian waves; hydrogen-
ion-potential recorders, which may be said to taste; thermometers;
pressure gauges of various sorts; microphones; and so on.   The
effectors may be electrical motors or solenoids or heating coils or
other instruments of very diverse sorts.   Between the receptor or
sense organ and the effector stands an intermediate set of elements,
whose function is to recombine the incoming impressions into such
form as to produce a desired type of response in the effectors.   The
information fed into this central control system will very often
contain information concerning the functioning of the effectors

themselves. These correspond among other things to the kinesthetic organs and other proprioceptors of the human system, for we too have organs which record the position of a joint or the rate of contraction of a muscle, etc. Moreover, the information received by the automaton need not be used at once but may be delayed or stored so as to become available at some future time. This is the analogue of memory. Finally, as long as the automaton is running, its very rules of operation are susceptible to some change on the basis of the data which have passed through its receptors in the past, and this is not unlike the process of learning.

The machines of which we are now speaking are not the dream of the sensationalist nor the hope of some future time. They already exist as thermostats, automatic gyrocompass ship-steering systems, self-propelled missiles—especially such as seek their target—anti-aircraft fire-control systems, automatically controlled oil-cracking stills, ultra-rapid computing machines, and the like. They had begun to be used long before the war—indeed, the very old steam-engine governor belongs among them—but the great mechanization of the Second World War brought them into their own, and the need of handling the extremely dangerous energy of the atom will probably bring them to a still higher point of development. Scarcely a month passes but a new book appears on these so-called control mechanisms, or servomechanisms, and the present age is as truly the age of servomechanisms as the nineteenth century was the age of the steam engine or the eighteenth century the age of the clock.

To sum up: the many automata of the present age are coupled to the outside world both for the reception of impressions and for the performance of actions. They contain sense organs, effectors, and the equivalent of a nervous system to integrate the transfer of information from the one to the other. They lend themselves very well to description in physiological terms. It is scarcely a miracle that they can be subsumed under one theory with the mechanisms of physiology.

The relation of these mechanisms to time demands careful study. It is clear, of course, that the relation input-output is a consecutive one in time and involves a definite past-future order. What is perhaps not so clear is that the theory of the sensitive automata is a statistical one. We are scarcely ever interested in the performance of a communication-engineering machine for a single input. To function adequately, it must give a satisfactory performance for a whole class of inputs, and this means a statistically satisfactory performance for the class of input which it is statistically expected to

receive.   Thus its theory belongs to the Gibbsian statistical mechanics
rather than to the classical Newtonian mechanics.   We shall study
this in much more detail in the chapter devoted to the theory of
communication.

Thus the modern automaton exists in the same sort of Bergsonian
time as the living organism; and hence there is no reason in Bergson's
considerations why the essential mode of functioning of the living
organism should not be the same as that of the automaton of this
type.   Vitalism has won to the extent that even mechanisms
correspond to the time-structure of vitalism; but as we have said,
this victory is a complete defeat, for from every point of view which
has the slightest relation to morality or religion, the new mechanics
is fully as mechanistic as the old.   Whether we should call the new
point of view materialistic is largely a question of words: the
ascendancy of matter characterizes a phase of nineteenth-century
physics far more than the present age, and "materialism" has come
to be but little more than a loose synonym for "mechanism."   In
fact, the whole mechanist-vitalist controversy has been relegated to
the limbo of badly posed questions.

II

Groups and Statistical Mechanics

At about the beginning of the present century, two scientists, one in the United States and one in France, were working along lines which would have seemed to each of them entirely unrelated, if either had had the remotest idea of the existence of the other. In New Haven, Willard Gibbs was developing his new point of view in statistical mechanics. In Paris, Henri Lebesgue was rivalling the fame of his master Emile Borel by the discovery of a revised and more powerful theory of integration for use in the study of trigonometric series. The two discoverers were alike in this, that each was a man of the study rather than of the laboratory, but from this point on, their whole attitudes to science were diametrically opposite.

Gibbs, mathematician though he was, always regarded mathematics as ancillary to physics. Lebesgue was an analyst of the purest type, an able exponent of the extremely exacting modern standards of mathematical rigor, and a writer whose works, as far as I know, do not contain one single example of a problem or a method originating directly from physics. Nevertheless, the work of these two men forms a single whole in which the questions asked by Gibbs find their answers, not in his own work but in the work of Lebesgue.

The key idea of Gibbs is this: in Newton's dynamics, in its original form, we are concerned with an individual system, with given initial velocities and momenta, undergoing changes according to a certain system of forces under the Newtonian laws which link force and acceleration. In the vast majority of practical cases, however, we are far from knowing all the initial velocities and momenta. If we assume a certain initial distribution of the incompletely known

45

positions and momenta of the system, this will determine in a completely Newtonian way the distribution of the momenta and positions for any future time. It will then be possible to make statements about these distributions, and some of these will have the character of assertions that the future system will have certain characteristics with probability one, or certain other characteristics with probability zero.

Probabilities one and zero are notions which include complete certainty and complete impossibility but include much more as well. If I shoot at a target with a bullet of the dimensions of a point, the chance that I hit any specific point on the target will generally be zero, although it is not impossible that I hit it; and indeed, in each specific case I must actually hit some specific point, which is an event of probability zero. Thus an event of probability one, that of my hitting *some* point, may be made up of an assemblage of instances of probability zero.

Nevertheless, one of the processes which is used in the technique of the Gibbsian statistical mechanics, although it is used implicitly, and Gibbs is nowhere clearly aware of it, is the resolution of a complex contingency into an infinite sequence of more special contingencies—a first, a second, a third, and so on—each of which has a known probability; and the expression of the probability of this larger contingency as the sum of the probabilities of the more special contingencies, which form an infinite sequence. Thus we *cannot* sum probabilities in all conceivable cases, to get a probability of the total event—for the sum of any number of zeros is zero—while we *can* sum them if there is a first, a second, a third member, and so on, forming a sequence of contingencies in which every term has a definite position given by a positive integer.

The distinction between these two cases involves rather subtle considerations concerning the nature of sets of instances, and Gibbs, although a very powerful mathematician, was never a very subtle one. Is it possible for a class to be infinite and yet essentially different in multiplicity from another infinite class, such as that of the positive integers? This problem was solved toward the end of the last century by Georg Cantor, and the answer is "Yes." If we consider all the distinct decimal fractions, terminating or nonterminating, lying between 0 and 1, it is known that they cannot be arranged in 1, 2, 3 order—although, strangely enough, all the *terminating* decimal fractions can be so arranged. Thus the distinction demanded by the Gibbs statistical mechanics is not on the face of it an impossible one. The service of Lebesgue to the Gibbs theory is to show that the implicit requirements of statistical mechanics concerning contingencies of probability zero and the addition of the probabilities of contingencies can actually be met, and that the Gibbsian theory does not involve contradictions.

Lebesgue's work, however, was not directly based on the needs of statistical mechanics but on what looks like a very different theory, the theory of trigonometric series. This goes back to the eighteenth-century physics of waves and vibrations, and to the then moot question of the generality of the sets of motions of a linear system which can be synthesized out of the simple vibrations of the system—out of those vibrations, in other words, for which the passing of time simply multiplies the deviations of the system from equilibrium by a quantity, positive or negative, dependent on the time alone and not on position. Thus a single function is expressed as the sum of a series. In these series, the coefficients are expressed as averages of the product of the function to be represented, multiplied by a given weighting function. The whole theory depends on the properties of the average of a series, in terms of the average of an individual term. Notice that the average of a quantity which is 1 over an interval from 0 to A, and 0 from A to 1, is A, and may be regarded as the probability that the random point should lie in the interval from 0 to A if it is known to lie between 0 and 1. In other words, the theory needed for the average of a series is very close to the theory needed for an adequate discussion of probabilities compounded from an infinite sequence of cases. This is the reason why Lebesgue, in solving his own problem, had also solved that of Gibbs.

The particular distributions discussed by Gibbs have themselves a dynamical interpretation. If we consider a certain very general sort of conservative dynamical system, with N degrees of freedom, we find that its position and velocity coordinates may be reduced to a special set of $2N$ coordinates, N of which are called the generalized position coordinates and N the generalized momenta. These determine a $2N$-dimensional space defining a $2N$-dimensional volume; and if we take any region of this space and let the points flow with the course of time, which changes every set of $2N$ coordinates into a new set depending on the elapsed time, the continual change of the boundary of the region does not change its $2N$-dimensional volume. In general, for sets not so simply defined as these regions, the notion of volume generates a system of measure of the type of Lebesgue. In this system of measure, and in the conservative dynamical systems which are transformed in such a way as to keep this measure constant, there is one other numerically valued entity which also remains constant: the energy. If all the bodies in the system act only on one another and there are no forces attached to fixed positions and fixed orientations in space, there are two other expressions which also remain constant. Both of these are vectors: the momentum, and the moment of momentum of the system as a whole. They are not difficult to eliminate, so that the system is replaced by a system with fewer degrees of freedom.

In highly specialized systems, there may be other quantities not determined by the energy, the momentum, and the moment of momentum, which are unchanged as the system develops. However, it is known that systems in which another invariant quantity exists, dependent on the initial coordinates and momenta of a dynamical system, and regular enough to be subject to the system of integration based on Lebesgue measure, are very rare indeed in a quite precise sense.[1] In systems without other invariant quantities, we can fix the coordinates corresponding to energy, momentum, and total moment of momentum, and in the space of the remaining coordinates, the measure determined by the position and momentum coordinates will itself determine a sort of sub-measure, just as measure in space will determine area on a two-dimensional surface out of a family of two-dimensional surfaces. For example, if our family is that of concentric spheres, then the volume between two concentric spheres close together, when normalized by taking as one the total volume of the region between the two spheres, will give in the limit a measure of area on the surface of a sphere.

Let us then take this new measure on a region in phase space for which energy, total momentum, and total moment of momentum are determined, and let us suppose that there are no other measurable invariant quantities in the system. Let the total measure of this restricted region be constant, or as as we can make it by a change in scale, 1. As our measure has been obtained from a measure invariant in time, in a way invariant in time, it will itself be invariant. We shall call this measure *phase measure*, and averages taken with respect to it *phase averages*.

However, any quantity varying in time may also have a *time average*. If, for example, $f(t)$ depends on t, its time average for the past will be

$$\lim_{T \to \infty} \frac{1}{T} \int_{-T}^{0} f(t)\, dt \qquad (2.01)$$

[1] Oxtoby, J. C., and S. M. Ulam,"Measure-Preserving Homeomorphisms and Metrical Transitivity." *Ann. of Math.*, Ser. 2, **42**, 874–920 (1941).

and its time average for the future

$$\lim_{T \to \infty} \frac{1}{T} \int_0^T f(t)\, dt \qquad (2.02)$$

In Gibbs' statistical mechanics, both time averages and space averages occur. It was a brilliant idea of Gibbs to try to show that these two types of average were, in some sense, the same. In the notion that these two types of average were related, Gibbs was perfectly right; and in the method by which he tried to show this relation, he was utterly and hopelessly wrong. For this he was scarcely to blame. Even at the time of his death, the fame of the Lebesgue integral had just begun to penetrate to America. For another fifteen years, it was a museum curiosity, only useful to show to young mathematicians the needs and possibilities of rigor. A mathematician as distinguished as W. F. Osgood[1] would have nothing to do with it till his dying day. It was not until about 1930 that a group of mathematicians—Koopman, von Neumann, Birkhoff [2] —finally established the proper foundations of the Gibbs statistical mechanics. Later, in the study of ergodic theory, we shall see what these foundations were.

Gibbs himself thought that in a system from which all the invariants had been removed as extra coordinates almost all paths of points in phase space passed through all coordinates in such a space. This hypothesis he called the *ergodic hypothesis*, from the Greek words ἔργον, "work," and ὁδός, "path." Now, in the first place, as Plancherel and others have shown, there is no significant case where that hypothesis is true. No differentiable path can cover an area in the plane, even if it is of infinite length. The followers of Gibbs, including at the end perhaps Gibbs himself, saw this in a vague way, and replaced this hypothesis by the *quasi-ergodic* hypothesis, which merely asserts that in the course of time a system generally passes indefinitely near to every point in the region of phase space determined by the known invariants. There is no logical difficulty as to the truth of this: it is merely quite inadequate for the conclusions which Gibbs bases on it. It says nothing about the relative time which the system spends in the neighborhood of each point.

Beside the notions of *average* and of *measure*—the average over a universe of a function 1 over a set to be measured and 0 elsewhere— which were most urgently needed to make sense out of Gibbs' theory,

---

[1] Nevertheless some of Osgood's early work represented an important step in the direction of the Lebesgue integral.

[2] Hopf, E., "Ergodentheorie," *Ergeb. Math.*, 5, No. 2, Springer, Berlin (1937).

in order to appreciate the real significance of ergodic theory we need a more precise analysis of the notion of *invariant*, as well as the notion of *transformation group*. These notions were certainly familiar to Gibbs, as his study of vector analysis shows. Nevertheless, it is possible to maintain that he did not assess them at their full philosophical value. Like his contemporary Heaviside, Gibbs is one of the scientists whose physico-mathematical acumen often outstrips their logic and who are generally right, while they are often unable to explain why and how they are right.

For the existence of any science, it is necessary that there exist phenomena which do not stand isolated. In a world ruled by a succession of miracles performed by an irrational God subject to sudden whims, we should be forced to await each new catastrophe in a state of perplexed passiveness. We have a picture of such a world in the croquet game in *Alice in Wonderland*; where the mallets are flamingos; the balls, hedgehogs, which quietly unroll and go about their own business; the hoops, playing-card soldiers, likewise subject to locomotor initiative of their own; and the rules are the decrees of the testy, unpredictable Queen of Hearts.

The essence of an effective rule for a game or a useful law of physics is that it be statable in advance, and that it apply to more than one case. Ideally, it should represent a property of the system discussed which remains the same under the flux of particular circumstances. In the simplest case, it is a property which is *invariant* to a set of *transformations* to which the system is subject. We are thus led to the notions of *transformation, transformation group*, and *invariant*.

A transformation of a system is some alteration in which each element goes into another. The modification of the solar system which occurs in the transition between time t_1 and time t_2 is a transformation of the sets of coordinates of the planets. The similar change in their coordinate when we move their origin, or subject our geometric axes to a rotation, is a transformation. The change in scale which occurs when we examine a preparation under the magnifying action of a microscope is likewise a transformation.

The result of following a transformation A by a transformation B is another transformation, known as the *product* or *resultant BA*. Note that in general it depends on the order of A and B. Thus if A is the transformation which takes the coordinate x into the coordinate y, and y into $-x$, while z is unchanged; while B takes x into z, z into $-x$, and y is unchanged; then BA will take x into y, y into $-z$, and z into $-x$; while AB will take x into z, y into $-x$, and z into $-y$. If

AB and BA are the same, we shall say that A and B are *permutable*.

Sometimes, but not always, the transformation A will not only carry every element of the system into an element but will have the property that every element is the result of transforming an element. In this case, there is a unique transformation A^{-1}, such that both AA^{-1} and $A^{-1}A$ are that very special transformation which we call I, the *identity transformation*, which transforms every element into itself. In this case we call A^{-1} the *inverse* of A. It is clear that A is the inverse of A^{-1}, that I is its own inverse, and that the inverse of AB is $B^{-1}A^{-1}$.

There exist certain sets of transformations where every transformation belonging to the set has an inverse, likewise belonging to the set; and where the resultant of any two transformations belonging to the set itself belongs to the set. These sets are known as *transformation groups*. The set of all translations along a line, or in a plane, or in a three-dimensional space, is a transformation group; and even more, it is a transformation group of the special sort known as *Abelian*, where any two transformations of the group are permutable. The set of rotations about a point, and of all motions of a rigid body in space, are non-Abelian groups.

Let us suppose that we have some quantity attached to all the elements transformed by a transformation group. If this quantity is unchanged when each element is changed by the same transformation of the group, whatever that transformation may be, it is called *an invariant of the group*. There are many sorts of such group invariants, of which two are especially important for our purposes.

The first are the so-called *linear invariants*. Let the elements transformed by an Abelian group be the terms which we represent by x, and let $f(x)$ be a complex-valued function of these elements, with certain appropriate properties of continuity or integrability. Then if Tx stands for the element resulting from x under the transformation T, and if $f(x)$ is a function of absolute value 1, such that

$$f(Tx) = \alpha(T)f(x) \qquad (2.03)$$

where $\alpha(T)$ is a number of absolute value 1 depending only on T, we shall say that $f(x)$ is a *character* of the group. It is an invariant of the group in a slightly generalized sense. If $f(x)$ and $g(x)$ are group characters, clearly $f(x)g(x)$ is one also, as is $[f(x)]^{-1}$. If we can represent any function $h(x)$ defined over the group as a linear combination of the characters of the group, in some such form as

$$h(x) = \sum A_k f_k(x) \qquad (2.04)$$

where $f_k(x)$ is a character of the group, and $\alpha_k(T)$ bears the same relation to $f_k(x)$ that $\alpha(T)$ does to $f(x)$ in Eq. 2.03, then

$$h(Tx) = \sum A_k \alpha_k(T) f_k(x) \qquad (2.05)$$

Thus if we can develop $h(x)$ in terms of a set of group characters, we can develop $h(Tx)$ for all T in terms of the characters.

We have seen that the characters of a group generate other characters under multiplication and inversion, and it may similarly be seen that the constant 1 is a character. Multiplication by a group character thus generates a transformation group of the group characters themselves, which is known as the *character group* of the original group.

If the original group is the translation group on the infinite line, so that the operator T changes x into $x + T$, Eq. 2.03 becomes

$$f(x + T) = \alpha(T) f(x) \qquad (2.06)$$

which is satisfied if $f(x) = e^{i\lambda x}$, $\alpha(T) = e^{i\lambda T}$. The characters will be the functions $e^{i\lambda x}$, and the character group will be the group of translations changing λ into $\lambda + \tau$, thus having the same structure as the original group. This will not be the case when the original group consists of the rotations about a circle. In this case, the operator T changes x into a number between 0 and 2π, differing from $x + T$ by an integral multiple of 2π, and, while Eq. 2.06 will still hold, we have the extra condition that

$$\alpha(T + 2\pi) = \alpha(T) \qquad (2.07)$$

If now we put $f(x) = e^{i\lambda x}$ as before, we shall obtain

$$e^{i 2\pi \lambda} = 1 \qquad (2.08)$$

which means that λ must be a real integer, positive, negative, or zero. The character group thus corresponds to the translations of the real integers. If, on the other hand, the original group is that of the translations of the integers, x and T in Eq. 2.05 are confined to the integer values, and $e^{i\lambda x}$ involves only the number between 0 and 2π which differs from λ by an integral multiple of 2π. Thus the character group is essentially the group of rotations about a circle.

In any character group, for a given character f, the values of $\alpha(T)$ are distributed in such a way that the distribution is not altered when they are all multiplied by $\alpha(S)$, for any element S in the group. That is, if there is any reasonable basis of taking an average of these values which is not affected by the transformation of the group by the multiplication of each transformation by a fixed one of its trans-formations, either $\alpha(T)$ is always 1, or this average is invariant when multiplied by some number not 1, and must be 0. From this it may be concluded that the average of the product of any character by its conjugate (which will also be a character) will have the value 1, and that the average of the product of any character by the conjugate of another character will have the value 0. In other words, if we can express $h(x)$ as in Eq. 2.04, we shall have

$$A_k = \text{average}\left[h(x)\overline{f_k(x)}\right] \tag{2.09}$$

In the case of the group of rotations on a circle, this gives us directly that if

$$f(x) = \sum a_n e^{inx} \tag{2.10}$$

then

$$a_n = \frac{1}{2\pi} \int_0^{2\pi} f(x)e^{-inx}\,dx \tag{2.11}$$

and the result for translations along the infinite line is closely related to the fact that if in an appropriate sense

$$f(x) = \int_{-\infty}^{\infty} a(\lambda)e^{i\lambda x}\,d\lambda \tag{2.12}$$

then in a certain sense

$$a(\lambda) = \frac{1}{2\pi} \int_{-\infty}^{\infty} f(x)e^{-i\lambda x}\,dx \tag{2.13}$$

These results have been stated here very roughly and without a clear statement of their conditions of validity. For more precise statements of the theory, the reader should consult the following reference.[1]

Beside the theory of the linear invariants of a group, there is also the general theory of its metrical invariants. These are the systems of Lebesgue measure which do not undergo any change when the objects transformed by the group are permuted by the operators of the group. In this connection, we should cite the interesting theory of group measure, due to Haar.[2] As we have seen, every group itself is a collection of objects which are permuted by being multiplied by the operations of the group itself. As such, it may have an invariant measure. Haar has proved that a certain rather wide class of

[1] Wiener, N., *The Fourier Integral and Certain of Its Applications*, The University Press, Cambridge, England, 1933; Dover Publications, Inc., N.Y.

[2] Haar, H., "Der Massbegriff in der Theorie der Kontinuierlichen Gruppen." *Ann. of Math.*, Ser. 2, **34**, 147–169 (1933).

groups does possess a uniquely determined invariant measure, definable in terms of the structure of the group itself.

The most important application of the theory of the metrical invariants of a group of transformations is to show the justification of that interchangeability of phase averages and time averages which, as we have already seen, Gibbs tried in vain to establish. The basis on which this has been accomplished is known as the ergodic theory.

The ordinary ergodic theorems start with an ensemble E, which we can take to be of measure 1, transformed into itself by a measure-preserving transformation T or by a group of measure-preserving transformations T^λ, where $-\infty < \lambda < \infty$ and where

$$T^\lambda \cdot T^\mu = T^{\lambda+\mu} \tag{2.14}$$

Ergodic theory concerns itself with complex-valued functions $f(x)$ of the elements x of E. In all cases, $f(x)$ is taken to be measurable in x, and if we are concerned with a continuous group of transformations, $f(T^\lambda x)$ is taken to be measurable in x and λ simultaneously.

In the mean ergodic theorem of Koopman and von Neumann, $f(x)$ is taken to be of class L^2; that is,

$$\int_E |f(x)|^2 \, dx < \infty \tag{2.15}$$

The theorem then asserts that

$$f_N(x) = \frac{1}{N+1} \sum_{n=0}^{N} f(T^n x) \tag{2.16}$$

or

$$f_A(x) = \frac{1}{A} \int_0^A f(T^\lambda x) \, d\lambda \tag{2.17}$$

as the case may be, converges in the mean to a limit $f^*(x)$ as $N \to \infty$ or $A \to \infty$, respectively, in the sense that

$$\lim_{N \to \infty} \int_E |f^*(x) - f_N(x)|^2 \, dx = 0 \tag{2.18}$$

$$\lim_{A \to \infty} \int_E |f^*(x) - f_A(x)|^2 \, dx = 0 \tag{2.19}$$

In the "almost everywhere" ergodic theorem of Birkhoff, $f(x)$ is taken to be of class L; which means that

$$\int_E |f(x)| \, dx < \infty \tag{2.20}$$

The functions $f_N(x)$ and $f_A(x)$ are defined as in Eqs. 2.16 and 2.17. The theorem then states that, except for a set of values of x of measure 0,

$$f^*(x) = \lim_{N \to \infty} f_N(x) \tag{2.21}$$

and

$$f^*(x) = \lim_{A \to \infty} f_A(x) \tag{2.22}$$

exist.

A very interesting case is the so-called *ergodic* or *metrically transitive* one, in which the transformation T or the set of transformations T^λ leaves invariant no set of points x which has a measure other than 1 or 0. In such a case, the set of values (for either ergodic theorem) for which f^* takes on a certain range of value is almost always either 1 or 0. This is impossible unless $f^*(x)$ is almost always constant. The value which $f^*(x)$ then assumes almost always is

$$\int_0^1 f(x)\, dx \tag{2.23}$$

That is, in the Koopman theorem, we have the limit in the mean

$$\text{l.i.m.} \frac{1}{N+1} \sum_{n=0}^N f(T^n x) = \int_0^1 f(x)\, dx \tag{2.24}$$

and in the Birkhoff theorem, we have

$$\lim_{N \to \infty} \frac{1}{N+1} \sum_{n=0}^N f(T^n x) = \int_0^1 f(x)\, dx \tag{2.25}$$

except for a set of values of x of zero measure or probability 0. Similar results hold in the continuous case. This is an adequate justification for Gibbs' interchange of phase averages and time averages.

Where the transformation T or the transformation group T^λ is not ergodic, von Neumann has shown under very general conditions that they can be reduced to ergodic components. That is, except for a set of values of x of zero measure, E can be separated into a finite or denumerable set of classes E_n and a continuum of classes $E(y)$, such that a measure is established on each E_n and $E(y)$, which is invariant under T or T^λ. These transformations are all ergodic; and if $S(y)$ is the intersection of S with $E(y)$ and S_n with E_n, then

$$\underset{E}{\text{measure}}\,(S) = \int \underset{E(y)}{\text{measure}}\,[S(y)]\, dy + \sum \underset{E_n}{\text{measure}}\,(S_n) \tag{2.26}$$

In other words, the whole theory of measure-preserving transformations can be reduced to the theory of ergodic transformations.

The whole of ergodic theory, let us remark in passing, may be applied to groups of transformations more general than those isomorphic with the translation group on the line. In particular, it may be applied to the translation group in n dimensions. The case of three dimensions is physically important. The spatial analogue of temporal equilibrium is spatial homogeneity, and such theories as that of the homogeneous gas, liquid, or solid depend on the application of three-dimensional ergodic theory. Incidentally, a non-ergodic group of translation transformations in three dimensions appears as the set of translations of a mixture of distinct states, such that one or another exists at a given time, not a mixture of both.

One of the cardinal notions of statistical mechanics, which also receives an application in the classical thermodynamics, is that of *entropy*. It is primarily a property of regions in phase space and expresses the logarithm of their probability measure. For example, let us consider the dynamics of n particles in a bottle, divided into two parts, A and B. If m particles are in A, and $n - m$ in B, we have characterized a region in phase space, and it will have a certain probability measure. The logarithm is the entropy of the distribution: m particles in A, $n - m$ in B. The system will spend most of its time in a state near that of greatest entropy, in the sense that for most of the time, nearly m_1 particles will be in A, nearly $n - m_1$ in B, where the probability of the combination m_1 in A, $n - m_1$ in B is a maximum. For systems with a large number of particles and states within the limits of practical discrimination, this means that if we take a state of other than maximum entropy and observe what happens to it, the entropy almost always increases.

In the ordinary thermodynamic problems of the heat engine, we are dealing with conditions in which we have a rough thermal equilibrium in large regions like an engine cylinder. The states for which we study the entropy are states involving maximum entropy for a given temperature and volume, for a small number of regions of the given volumes and at the given temperature assumed. Even the more refined discussions of thermal engines, particularly of thermal engines like the turbine, in which a gas is expanding in a more complicated manner than in a cylinder, do not change these conditions too radically. We may still talk of local temperatures, with a very fair approximation, even though no temperature is precisely determined except in a state of equilibrium and by methods involving this equilibrium. However, in living matter, we lose much of even this rough homogeneity. The structure of a protein tissue as shown by the electron microscope has an enormous definiteness and fineness of texture, and its physiology is certainly of a corresponding fineness of texture. This fineness is far greater than that of the space-and-time scale of an ordinary thermometer, and so the temperatures read by ordinary thermometers in living tissues are gross averages and not the true temperatures of thermodynamics. Gibbsian statistical mechanics may well be a fairly adequate model of what happens in the body; the picture suggested by the ordinary heat engine certainly is not. The thermal efficiency of muscle action means next to nothing, and certainly does not mean what it appears to mean.

A very important idea in statistical mechanics is that of the Maxwell demon. Let us suppose a gas in which the particles are moving around with the distribution of velocities in statistical equilibrium for a given temperature. For a perfect gas, this is the Maxwell distribution. Let this gas be contained in a rigid container with a wall across it, containing an opening spanned by a small gate, operated by a gatekeeper, either an anthropomorphic demon or a minute mechanism. When a particle of more than average velocity approaches the gate from compartment A or a particle of less than average velocity approaches the gate from compartment B, the gatekeeper opens the gate, and the particle passes through; but when a particle of less than average velocity approaches from compartment A or a particle of greater than average velocity approaches from compartment B, the gate is closed. In this way, the concentration of particles of high velocity is increased in compartment B and is decreased in compartment A. This produces an apparent decrease in entropy; so that if the two compartments are now connected by a heat engine, we seem to obtain a perpetual-motion machine of the second kind.

It is simpler to repel the question posed by the Maxwell demon than to answer it. Nothing is easier than to deny the possibility of such beings or structures. We shall actually find that Maxwell demons in the strictest sense cannot exist in a system in equilibrium, but if we accept this from the beginning, and so not try to demonstrate it, we shall miss an admirable opportunity to learn something about entropy and about possible physical, chemical, and biological systems.

For a Maxwell demon to act, it must receive information from approaching particles concerning their velocity and point of impact on the wall. Whether these impulses involve a transfer of energy or not, they must involve a coupling of the demon and the gas. Now, the law of the increase of entropy applies to a completely isolated system but does not apply to a non-isolated part of such a system. Accordingly, the only entropy which concerns us is that of the system gas-demon, and not that of the gas alone. The gas entropy is merely one term in the total entropy of the larger system. Can we find terms involving the demon as well which contribute to this total entropy?

Most certainly we can. The demon can only act on information received, and this information, as we shall see in the next chapter, represents a negative entropy. The information must be carried by some physical process, say some form of radiation. It may very well be that this information is carried at a very low energy level, and that the transfer of energy between particle and demon is for a considerable time far less significant than the transfer of information. However, under the quantum mechanics, it is impossible to obtain any information giving the position or the momentum of a particle, much less the two together, without a positive effect on the energy of the particle examined, exceeding a minimum dependent on the frequency of the light used for examination. Thus all coupling is strictly a coupling involving energy, and a system in statistical equilibrium is in equilibrium both in matters concerning entropy and those concerning energy. In the long run, the Maxwell demon is itself subject to a random motion corresponding to the temperature of its environment, and, as Leibniz says of some of his monads, it receives a large number of small impressions, until it falls into "a certain vertigo" and is incapable of clear perceptions. In fact, it ceases to act as a Maxwell demon.

Nevertheless, there may be a quite appreciable interval of time before the demon is deconditioned, and this time may be so prolonged that we may speak of the active phase of the demon as metastable. There is no reason to suppose that metastable demons do not in fact exist; indeed, it may well be that enzymes are metastable Maxwell demons, decreasing entropy, perhaps not by the separation between fast and slow particles but by some other equivalent process. We may well regard living organisms, such as Man himself, in this light. Certainly the enzyme and the living organism are alike metastable: the stable state of an enzyme is to be deconditioned, and the stable state of a living organism is to be dead. All catalysts are ultimately poisoned: they change rates of reaction but not true equilibrium. Nevertheless, catalysts and Man alike have sufficiently definite states of metastability to deserve the recognition of these states as relatively permanent conditions.

I do not wish to close this chapter without indicating that ergodic theory is a considerably wider subject than we have indicated above. There are certain modern developments of ergodic theory in which the measure to be kept invariant under a set of transformations is defined directly by the set itself rather than assumed in advance. I refer especially to the work of Kryloff and Bogoliouboff, and to some of the work of Hurewicz and the Japanese school.

The next chapter is devoted to the statistical mechanics of time series. This is another field in which the conditions are very remote from those of the statistical mechanics of heat engines and which is thus very well suited to serve as a model of what happens in the living organism.

III

Time Series, Information, and Communication

There is a large class of phenomena in which what is observed is a numerical quantity, or a sequence of numerical quantities, distributed in time. The temperature as recorded by a continuous recording thermometer, or the closing quotations of a stock in the stock market, taken day by day, or the complete set of meteorological data published from day to day by the Weather Bureau are all time series, continuous or discrete, simple or multiple. These time series are relatively slowly changing, and are well suited to a treatment employing hand computation or ordinary numerical tools such as slide rules and computing machines. Their study belongs to the more conventional parts of statistical theory.

What is not generally realized is that the rapidly changing sequences of voltages in a telephone line or a television circuit or a piece of radar apparatus belong just as truly to the field of statistics and time series, although the apparatus by means of which they are combined and modified must in general be very rapid in its action, and in fact must be able to put out results *pari passu* with the very rapid alterations of input. These pieces of apparatus—telephone receivers, wave filters, automatic sound-coding devices like the Vocoder of the Bell Telephone Laboratories, frequency-modulating networks and their corresponding receivers—are all in essence quick-acting arithmetical devices, corresponding to the whole apparatus of computing machines and schedules, and the staff of computers, of the statistical laboratory. The ingenuity needed in their use has been built into them in advance, just as it has into the automatic range finders and gun pointers of an anti-aircraft fire-control system, and for the same reasons. The chain of operation has to work too fast to admit of any human links.

One and all, time series and the apparatus to deal with them, whether in the computing laboratory or in the telephone circuit, have to deal with the recording, preservation, transmission, and use of information. What is this information, and how is it measured? One of the simplest, most unitary forms of information is the recording of a choice between two equally probable simple alternatives, one or the other of which is bound to happen—a choice, for example, between heads and tails in the tossing of a coin. We shall call a single choice of this sort a *decision*. If then we ask for the amount of information in the perfectly precise measurement of a quantity known to lie between A and B, which may with uniform *a priori* probability lie anywhere in this range, we shall see that if we put $A = 0$ and $B = 1$, and represent the quantity in the binary scale by the infinite binary number $. a_1 a_2 a_3 \cdots a_n \cdots$, where $a_1, a_2, \cdots$, each has the value 0 or 1, then the number of choices made and the consequent amount of information is infinite. Here

$$. a_1 a_2 a_3 \cdots a_n \cdots = \frac{1}{2} a_1 + \frac{1}{2^2} a_2 + \cdots + \frac{1}{2^n} a_n + \cdots \quad (3.01)$$

However, no measurement which we actually make is performed with perfect precision. If the measurement has a uniformly distributed error lying over a range of length $\cdot b_1 b_2 \cdots b_n \cdots$, where b_k is the first digit not equal to 0, we shall see that all the decisions from a_1 to a_{k-1}, and possibly to a_k, are significant, while all the later decisions are not. The number of decisions made is certainly not far from

$$-\log_2 . b_1 b_2 \cdots b_n \cdots \quad (3.02)$$

and we shall take this quantity as the precise formula for the amount of information and its definition.

We may conceive this in the following way: we know *a priori* that a variable lies between 0 and 1, and *a posteriori* that it lies on the interval (a, b) inside $(0, 1)$. Then the amount of information we have from our *a posteriori* knowledge is

$$-\log_2 \frac{\text{measure of } (a, b)}{\text{measure of } (0, 1)} \quad (3.03)$$

However, let us now consider a case where our *a priori* knowledge is that the probability that a certain quantity should lie between x and

$x + dx$ is $f_1(x)\, dx$, and the *a posteriori* probability is $f_2(x)\, dx$. How much new information does our *a posteriori* probability give us?

This problem is essentially that of attaching a width to the regions under the curves $y = f_1(x)$ and $y = f_2(x)$. It will be noted that we are here assuming the variable to have a fundamental equipartition; that is, our results will not in general be the same if we replace x by x^3 or any other function of x. Since $f_1(x)$ is a probability density, we shall have

$$\int_{-\infty}^{\infty} f_1(x)\, dx = 1 \qquad (3.04)$$

so that the average logarithm of the breadth of the region under $f_1(x)$ may be considered as some sort of average of the height of the logarithm of the reciprocal of $f_1(x)$. Thus a reasonable measure[1] of the amount of information associated with the curve $f_1(x)$ is

$$\int_{-\infty}^{\infty} [\log_2 f_1(x)] f_1(x)\, dx \qquad (3.05)$$

The quantity we here define as amount of information is the negative of the quantity usually defined as entropy in similar situations. The definition here given is not the one given by R. A. Fisher for statistical problems, although it is a statistical definition; and can be used to replace Fisher's definition in the technique of statistics.

In particular, if $f_1(x)$ is a constant over (a, b) and is zero elsewhere,

$$\int_{-\infty}^{\infty} [\log_2 f_1(x)] f_1(x)\, dx = \frac{b-a}{b-a} \log_2 \frac{1}{b-a} = \log_2 \frac{1}{b-a} \qquad (3.06)$$

Using this to compare the information that a point is in the region $(0, 1)$ with the information that it is in the region (a, b), we obtain for the measure of the difference

$$\log_2 \frac{1}{b-a} - \log_2 1 = \log_2 \frac{1}{b-a} \qquad (3.07)$$

The definition which we have given for the amount of information is applicable when the variable x is replaced by a variable ranging over two or more dimensions. In the two-dimensional case, $f(x, y)$ is a function such that

$$\int_{-\infty}^{\infty} dx \int_{-\infty}^{\infty} dy\, f_1(x, y) = 1 \qquad (3.08)$$

[1] Here the author makes use of a personal communication of J. von Neumann.

and the amount of information is

$$\int_{-\infty}^{\infty} dx \int_{-\infty}^{\infty} dy \, f_1(x, y) \log_2 f_1(x, y) \tag{3.081}$$

Note that if $f_1(x, y)$ is of the form $\phi(x)\psi(y)$ and

$$\int_{-\infty}^{\infty} \phi(x) \, dx = \int_{-\infty}^{\infty} \psi(y) \, dy = 1 \tag{3.082}$$

then

$$\int_{-\infty}^{\infty} dx \int_{-\infty}^{\infty} dy \, \phi(x)\psi(y) = 1 \tag{3.083}$$

and

$$\int_{-\infty}^{\infty} dx \int_{-\infty}^{\infty} dy \, f_1(x, y) \log_2 f_1(x, y)$$
$$= \int_{-\infty}^{\infty} dx \, \phi(x) \log_2 \phi(x) + \int_{-\infty}^{\infty} dy \, \psi(y) \log_2 \psi(y) \tag{3.084}$$

and the amount of information from independent sources is additive.

An interesting problem is that of determining the information gained by fixing one or more variables in a problem. For example, let us suppose that a variable u lies between x and $x + dx$ with the probability $\exp(-x^2/2a) \, dx/\sqrt{2\pi a}$, while a variable v lies between the same two limits with a probability $\exp(-x^2/2b) \, dx/\sqrt{2\pi b}$. How much information do we gain concerning u if we know that $u + v = w$? In this case, it is clear that $u = w - v$, where w is fixed. We assume the *a priori* distributions of u and v to be independent. Then the *a posteriori* distribution of u is proportional to

$$\exp\left(-\frac{x^2}{2a}\right) \exp\left[-\frac{(w-x)^2}{2b}\right] = c_1 \exp\left[-(x - c_2)^2 \left(\frac{a+b}{2ab}\right)\right] \tag{3.09}$$

where c_1 and c_2 are constants. They both disappear in the formula for the gain in information given by the fixing of w.

The excess of information concerning x when we know w to be that which we have in advance is

$$\frac{1}{\sqrt{2\pi[ab/(a+b)]}} \int_{-\infty}^{\infty} \left\{ \exp\left[-(x - c_2)^2\left(\frac{a+b}{2ab}\right)\right] \right\}$$
$$\times \left[-\frac{1}{2}\log_2 2\pi\left(\frac{ab}{a+b}\right)\right] - (x - c_2)^2 \left[\left(\frac{a+b}{2ab}\right)\right] \log_2 e \right] dx$$
$$-\frac{1}{\sqrt{2\pi a}} \int_{-\infty}^{\infty} \left[\exp\left(-\frac{x^2}{2a}\right)\right]\left(-\frac{1}{2}\log_2 2\pi a - \frac{x^2}{2a}\log_2 e\right) dx$$
$$= \frac{1}{2} \log_2\left(\frac{a+b}{b}\right) \tag{3.091}$$

Note that this expression (Eq. 3.091) is positive, and that it is independent of w. It is one-half the logarithm of the ratio of the sum of the mean squares of u and v to the mean square of v. If v has only a small range of variation, the amount of information concerning u which a knowledge of $u + v$ gives is large, and it becomes infinite as b goes to 0.

We can consider this result in the following light: let us treat u as a message and v as a noise. Then the information carried by a precise message in the absence of a noise is infinite. In the presence of a noise, however, this amount of information is finite, and it approaches 0 very rapidly as the noise increases in intensity.

We have said that amount of information, being the negative logarithm of a quantity which we may consider as a probability, is essentially a negative entropy. It is interesting to show that, on the average, it has the properties we associate with an entropy.

Let $\phi(x)$ and $\psi(x)$ be two probability densities; then $[\phi(x) + \psi(x)]/2$ is also a probability density. Then

$$\int_{-\infty}^{\infty} \frac{\phi(x) + \psi(x)}{2} \log \frac{\phi(x) + \psi(x)}{2}\, dx$$

$$\leqslant \int_{-\infty}^{\infty} \frac{\phi(x)}{2} \log \phi(x)\, dx + \int_{-\infty}^{\infty} \frac{\psi(x)}{2} \log \psi(x)\, dx \quad (3.10)$$

This follows from the fact that

$$\frac{a + b}{2} \log \frac{a + b}{2} \leqslant \frac{1}{2}\,(a \log a + b \log b) \quad (3.11)$$

In other words, the overlap of the regions under $\phi(x)$ and $\psi(x)$ reduces the maximum information belonging to $\phi(x) + \psi(x)$. On the other hand, if $\phi(x)$ is a probability density vanishing outside (a, b),

$$\int_{-\infty}^{\infty} \phi(x) \log \phi(x)\, dx \quad (3.12)$$

is a minimum when $\phi(x) = 1/(b - a)$ over (a, b) and is zero elsewhere. This follows from the fact that the logarithm curve is convex upward.

It will be seen that the processes which lose information are, as we should expect, closely analogous to the processes which gain entropy. They consist in the fusion of regions of probability which were originally distinct. For example, if we replace the distribution of a certain variable by the distribution of a function of that variable which takes the same value for different arguments, or if in a function of several variables we allow some of them to range unimpeded over their natural range of variability, we lose information.   No operation on a message can gain information on the average.   Here we have a precise application of the second law of thermodynamics in communication engineering.   Conversely, the greater specification of an ambiguous situation, on the average, will, as we have seen, generally gain information and never lose it.

An interesting case is when we have a probability distribution with n-fold density $f(x_1, \cdots, x_n)$ over the variables $(x_1, \cdots, x_n)$, and where we have m dependent variables $y_1, \cdots, y_m$.   How much information do we get by fixing these m variables?   First let them be fixed between the limits $y_1^*, y_1^* + dy_1^*; \cdots; y_m^*, y_m + dy_m^*$.   Let us take as a new set of variables $x_1, x_2, \cdots, x_{n-m}, y_1, y_2, \cdots, y_m$. Then over the new set of variables, our distribution function will be proportional to $f(x_1, \cdots, x_n)$ over the region R given by $y_1^* \leqslant y_1 \leqslant y_1^* + dy_1^*, \cdots, y_m^* \leqslant y_m \leqslant y_m^* + dy_m^*$ and 0 outside. Thus the amount of information obtained by the specification of the y's will be

$$\frac{\underbrace{\int dx_1 \cdots \int dx_n\, f(x_1, \cdots, x_n)\, \log_2 f(x_1, \cdots, x_n)}_{R}}{\underbrace{\int dx_1 \cdots \int dx_n\, f(x_1, \cdots, x_n)}_{R}}$$

$$= \left\{ \begin{array}{c} -\displaystyle\int_{-\infty}^{\infty} dx_1 \cdots \int_{-\infty}^{\infty} dx_n\, f(x_1, \cdots, x_n)\, \log_2 f(x_1, \cdots, x_n) \\[2ex] \dfrac{\displaystyle\int_{-\infty}^{\infty} dx_1 \cdots \int_{-\infty}^{\infty} dx_{n-m} \left| J\!\left(\begin{array}{c} y_1^*, \cdots, y_m^* \\ x_{n-m+1}, \cdots, x_n \end{array} \right) \right|^{-1} \times f(x_1, \cdots, x_n)\, \log_2 f(x_1, \cdots, x_n)}{\displaystyle\int_{-\infty}^{\infty} dx_1 \cdots \int_{-\infty}^{\infty} dx_{n-m} \left| J\!\left(\begin{array}{c} y_1^*, \cdots, y_m^* \\ x_{n-m+1}, \cdots, x_n \end{array} \right) \right|^{-1} f(x_1, \cdots, x_n)} \\[4ex] -\displaystyle\int_{-\infty}^{\infty} dx_1 \cdots \int_{-\infty}^{\infty} dx_n\, f(x_1, \cdots, x_n)\, \log_2 f(x_1, \cdots, x_n) \end{array} \right\}$$

$$(3.13)$$

Closely related to this problem is the generalization of that which we discussed in Eq. 3.13; in the case just discussed, how much information do we have concerning the variables $x_1, \cdots, x_{n-m}$ alone? Here the *a priori* probability density of these variables is

$$\int_{-\infty}^{\infty} dx_{n-m+1} \cdots \int_{-\infty}^{\infty} dx_n\, f(x_1, \cdots, x_n) \qquad (3.14)$$

and the un-normalized probability density after fixing the y^*'s is

$$\sum \left| J\begin{pmatrix} y_1^*, & \cdots, & y_m^* \\ x_{n-m+1}, & \cdots, & x_n \end{pmatrix} \right|^{-1} f(x_1, \cdots, x_n) \qquad (3.141)$$

where the $\sum$ is taken over all sets of points $(x_{n-m+1}, \cdots, x_n)$ corresponding to a given set of y^*'s. On this basis, we may easily write down the solution to our problem, though it will be a bit lengthy. If we take the set $x_1, \cdots, x_{n-m})$ to be a generalized message, the set $(x_{n-m+1}, \cdots, x_m)$ to be a generalized noise, and the y^*'s to be a generalized corrupted message, we see that we have given the solution of a generalization of the problem of Expression 3.141.

We have thus at least a formal solution of a generalization of the message-noise problem which we have already stated. A set of observations depends in an arbitrary way on a set of messages and noises with a known combined distribution. We wish to ascertain how much information these observations give concerning the messages alone. This is a central problem of communication engineering. It enables us to evaluate different systems, such as amplitude modulation or frequency modulation or phase modulation, as far as their efficiency in transmitting information is concerned. This is a technical problem and not suitable to a detailed discussion here; however, certain remarks are in order. In the first place, it can be shown that with the definition of information given here, with a random "static" on the ether equidistributed in frequency as far as power is concerned, and with a message restricted to a definite frequency range and a definite power output for this range, no means of transmission of information is more efficient than amplitude modulation, although other means may be as efficient. On the other hand, the information transmitted by this means is not necessarily in the form most suitable for reception by the ear or by any other given receptor. Here the specific characteristics of the ear and of other receptors must be considered by employing a theory very similar to the one just developed. In general, the efficient use of amplitude modulation or any other form of modulation must be supplemented by the use of decoding devices adequate to transforming the received information into a form suitable for reception by human receptors or

use by mechanical receptors.   Similarly, the original message must be
coded for the greatest compression in transmission.   This problem
has been attacked, at least in part, in the design of the "Vocoder"
system by the Bell Telephone Laboratories, and the relevant general
theory has been presented in a very satisfactory form by Dr. C.
Shannon of those laboratories.

So much for the definition and technique of measuring information.
We shall now discuss the way in which information may be presented
in a form homogeneous in time.   Let it be noted that most of the
telephone and other communication devices are actually not attached
to a particular origin in time.   There is indeed one operation which
seems to contradict this, but which really does not.   This is the
operation of modulation.   This, in its simplest form, converts a
message $f(t)$ into one of the form $f(t) \sin (at + b)$.   If, however, we
regard the factor $\sin (at + b)$ as an extra message which is put into
the apparatus, it will be seen that the situation will come under our
general statement.   The extra message, which we call the *carrier*,
adds nothing to the rate at which the system is carrying information.
All the information it contains is conveyed in an arbitrarily short
interval of time, and thereafter nothing new is said.

A message homogeneous in time, or, as the statisticians call it, a
time series which is in statistical equilibrium, is thus a single function
or a set of functions of the time, which forms one of an ensemble of
such sets with a well-defined probability distribution, not altered by
the change of t to $t + \tau$ throughout.   That is, the transformation
group consisting of the operators T^λ which change $f(t)$ into $f(t + \lambda)$
leaves the probability of the ensemble invariant.   The group
satisfies the properties that

$$T^\lambda[T^\mu f(t)] = T^{\mu+\lambda}f(t) \qquad \begin{cases} (-\infty < \lambda < \infty) \\ (-\infty < \mu < \infty) \end{cases} \qquad (3.15)$$

It follows from this that if $\Phi[f(t)]$ is a "functional" of $f(t)$—that is,
a number depending upon the whole history of $f(t)$—and if the
average of $f(t)$ over the whole ensemble is finite, we are in a position
to use the Birkhoff ergodic theorem quoted in the previous chapter,
and to come to the conclusion that, except for a set of values of
$f(t)$ of zero probability, the time-average of $\Phi[f(t)]$, or in symbols,

$$\lim_{A\to\infty} \frac{1}{A} \int_0^A \Phi[f(t + \tau)] \, d\tau = \lim_{A\to\infty} \frac{1}{A} \int_{-A}^0 \Phi[f(t + \tau)] \, d\tau \quad (3.16)$$

exists.

There is even more here than this.   We have stated in the previous chapter another theorem of ergodic character, due to von Neumann, which states that, except for a set of elements of zero probability, any element belonging to a system which goes into itself under a group of measure-preserving transformations such as Eq. 3.15 belongs to a sub-set (which may be the whole set) which goes into itself under the same transformation, which has a measure defined over itself and also invariant under the transformation, and which has the further property that any portion of this sub-set with measure preserved under the group of transformations either has the maximum measure of the sub-set, or measure 0. If we discard all elements except those of such a sub-set, and use its appropriate measure, we shall find that the time average (Eq. 3.16) is in almost all cases the average of $\Phi[f(t)]$ over all the space of functions $f(t)$; the so-called *phase average*. Thus in the case of such an ensemble of functions $f(t)$, except in a set of cases of zero probability, we can deduce the average of any statistical parameter of the ensemble—indeed we can simultaneously deduce any countable set of such parameters of the ensemble—from the record of any one of the component time series, by using a time average instead of a phase average. Moreover, we need to know only the past of almost any one time series of the class. In other words, given the entire history up to the present of a time series known to belong to an ensemble in statistical equilibrium, we can compute with probable error zero the entire set of statistical parameters of an ensemble in statistical equilibrium to which that time series belongs. Up to here, we have formulated this for single time series; it is equally true, however, for multiple time series in which we have several quantities varying simultaneously, rather than a single varying quantity.

We are now in a position to discuss various problems belonging to time series. We shall confine our attention to those cases where the entire past of a time series can be given in terms of a countable set of quantities. For example, for quite a wide class of functions $f(t)$ $(-\infty < t < \infty)$, we have fully determined f when we know the set of quantities

$$a_n = \int_{-\infty}^{0} e^{itn} f(t)\, dt \qquad (n = 0, 1, 2, \cdots) \qquad (3.17)$$

Now let A be some function of the values of t in the future, that is, for arguments greater than 0. Then we can determine the simultaneous distribution of $(a_0, a_1, \cdots, a_n, A)$ from the past of almost any single time series if the set of f's is taken in its narrowest possible sense. In particular, if $a_0, \cdots, a_n$ are all given, we may determine the distribution of A. Here we appeal to the known theorem of Nikodym on conditional probabilities. The same theorem will assure us that this distribution, under very general circumstances, will tend to a limit as $n \to \infty$ and this limit will give us all the knowledge there is concerning the distribution of any future quantity. We may similarly determine the simultaneous distribution of values of any set of future quantities, or any set of quantities depending both on the past and the future, when the past is known. If then we have given any adequate interpretation to the "*best* value" of any of these statistical parameters or sets of statistical parameters— in the sense, perhaps, of a mean or a median or a mode—we can compute it from the known distribution, and obtain a prediction to meet any desired criterion of goodness of prediction. We can compute the merit of the prediction, using any desired statistical basis of this merit—mean square error or maximum error or mean absolute error, and so on. We can compute the amount of information concerning any statistical parameter or set of statistical parameters, which fixing of the past will give us. We can even compute the whole amount of information which a knowledge of the past will give us of the whole future beyond a certain point; although when this point is the present, we shall in general know the latter from the past, and our knowledge of the present will contain an infinite amount of information.

Another interesting situation is that of a multiple time series, in which we know precisely only the pasts of some of the components. The distribution of any quantity involving more than these pasts can be studied by means very similar to those already suggested. In particular, we may wish to know the distribution of a value of another component, or a set of values of other components, at some point of time, past, present, or future. The general problem of the wave filter belongs to this class. We have a message, together with a noise, combined in some way into a corrupted message, of which we know the past. We also know the statistical joint distribution of the message and the noise as time series. We ask for the distribution of the values of the message at some given time, past, present, and future. We then ask for an operator on the past of the corrupted message which will *best* give this true message, in some given statistical sense. We may ask for a statistical estimate of some measure of the error of our knowledge of the message. Finally, we may ask for the amount of information which we possess concerning the message.

There is one ensemble of time series which is particularly simple and

central.   This is the ensemble associated with the Brownian motion.
The Brownian motion is the motion of a particle in a gas, impelled by
the random impacts of the other particles in a state of thermal
agitation.   The theory has been developed by many writers, among
them Einstein, Smoluchowski, Perrin, and the author.[1]   Unless we
go down in the time scale to intervals so small that the individual
impacts of the particles on one another are discernible, the motion
shows a curious kind of undifferentiability.   The mean square
motion in a given direction over a given time is proportional to the
length of that time, and the motions over successive times are
completely uncorrelated.   This conforms closely to the physical
observations.   If we normalize the scale of the Brownian motion to
fit the time scale, and consider only one coordinate x of the motion,
and if we let $x(t)$ equal 0 for $t = 0$, then the probability that if
$0 \leqslant t_1 \leqslant t_2 \leqslant \cdots \leqslant t_n$ the particles lie between x_1 and $x_1 + dx_1$ at
time $t_1, \cdots$, between x_n and $x_n + dx_n$ at time t_n, is

$$\frac{\exp\left[-\dfrac{x_1{}^2}{2t_1} - \dfrac{(x_2 - x_1)^2}{2(t_2 - t_1)} - \cdots - \dfrac{(x_n - x_{n-1})^2}{2(t_n - t_{n-1})}\right]}{\sqrt{|(2\pi)^n t_1(t_2 - t_1) \cdots (t_n - t_{n-1})|}} \, dx_1 \cdots dx_n \quad (3.18)$$

On the basis of the probability system corresponding to this,
which is unambiguous, we can make the set of paths corresponding
to the different possible Brownian motions depend on a parameter α
lying between 0 and 1, in such a way that each path is a function
$x(t, \alpha)$, where x depends on the time t and the parameter of distribution α, and where the probability that a path lies in a certain set S
is the same as the measure of the set of values of α corresponding to
paths in S.   On this basis, almost all paths will be continuous and
non-differentiable.

A very interesting question is that of determining the average with
respect to α of $x(t_1, \alpha) \cdots x(t_n, \alpha)$.   This will be

$$\int_0^1 d\alpha \, x(t_1, \alpha) x(t_2, \alpha) \cdots x(t_n, \alpha)$$

$$= (2\pi)^{-n/2} [t_1(t_2 - t_1) \cdots (t_n - t_{n-1})]^{-1/2}$$

$$\times \int_{-\infty}^{\infty} d\xi_1 \cdots \int_{-\infty}^{\infty} d\xi_n \, \xi_1 \xi_2 \cdots \xi_n \exp\left[-\frac{\xi_1{}^2}{2t_1} - \frac{(\xi_2 - \xi_1)^2}{2(t_2 - t_1)} - \cdots \right.$$

$$\left. - \frac{(\xi_n - \xi_{n-1})^2}{2(t_n - t_{n-1})}\right] \quad (3.19)$$

---

[1] Paley, R. E. A. C., and N. Wiener, "Fourier Transforms in the Complex Domain,"
Colloquium Publications, Vol. 19, American Mathematical Society, New York, 1934,
Chapter 10.

under the assumption that $0 \leqslant t_1 \leqslant \cdots \leqslant t_n$. Let us put

$$\xi_1 \cdots \xi_n = \sum A_k \xi_1^{\lambda_{k,1}} (\xi_2 - \xi_1)^{\lambda_{k,2}} \cdots (\xi_n - \xi_{n-1})^{\lambda_{k,n}} \qquad (3.20)$$

where $\lambda_{k,1} + \lambda_{k,2} + \cdots + \lambda_{k,n} = n$. The value of the expression in Eq. 3.19 will become

$$\sum A_k (2\pi)^{-n/2} [t_1^{\lambda_{k,1}} (t_2 - t_1)^{\lambda_{k,2}} \cdots (t_n - t_{n-1})^{\lambda_{k,n}}]^{-1/2}$$

$$\times \prod_j \int_{-\infty}^{\infty} d\xi \; \xi^{\lambda_{k,j}} \exp \left[-\frac{\xi^2}{2(t_j - t_{j-1})} \right]$$

$$= \sum A_k \prod_j \frac{1}{\sqrt{2\pi}} \int_{-\infty}^{\infty} \xi^{\lambda_{k,j}} \exp \left(-\frac{\xi^2}{2} \right) d\xi (t_j - t_{j-1})^{-1/2}$$

$$= \begin{cases} 0 \text{ if any } \lambda_{k,j} \text{ is odd} \\ \sum_k A_k \prod_j (\lambda_{k,j} - 1)(\lambda_{k,j} - 3) \cdots 5 \cdot 3 \cdot (t_j - t_{j-1})^{-1/2} \end{cases} \qquad (3.21)$$

if every $\lambda_{k,j}$ is even,

$$= \sum_k A_k \prod_j \text{(number of ways of dividing } \lambda_{k,j} \text{ terms into pairs)}$$
$$\times (t_j - t_{j-1})^{1/2}$$

$$= \sum_k A_k \text{ (numbers of ways of dividing } n \text{ terms into pairs}$$
$$\text{whose elements both belong in the same group of}$$
$$\lambda_{k,j} \text{ terms into which } \lambda \text{ is separated)} \times (t_j - t_{j-1})^{1/2}$$

$$= \sum_j A_j \sum \prod \int_0^1 d\alpha \, [x(t_k, \alpha) - x(t_{k-1}, \alpha)][x(t_q, \alpha) - x(t_{q-1}, \alpha)]$$

Here the first $\sum$ sums over j; the second, over all the ways of dividing n terms in blocks, respectively, of $\lambda_{k,1}, \cdots, \lambda_{k,n}$ numbers into pairs; and the $\prod$ is taken over those pairs of values k and q, where $\lambda_{k,1}$ of the elements to be selected from t_k and t_q are t_1, $\lambda_{k,2}$ are t_2, and so on. It immediately results that

$$\int_0^1 d\alpha \, x(t_1, \alpha) x(t_2, \alpha) \cdots x(t_n, \alpha) = \sum \prod \int_0^1 d\alpha \, x(t_j, \alpha) x(t_k, \alpha) \qquad (3.22)$$

where the $\sum$ is taken over all partitions of $t_1, \cdots, t_n$ into distinct pairs, and the $\prod$ over all the pairs in each partition. In other words, when we know the averages of the products of $x(t_j, \alpha)$ by pairs, we know the averages of all polynomials in these quantities, and thus their entire statistical distribution.

Up to the present, we have considered Brownian motions $x(t, \alpha)$ where t is positive. If we put

$$\begin{aligned} \xi(t, \alpha, \beta) &= x(t, \alpha) & (t \geqslant 0) \\ \xi(t, \alpha, \beta) &= x(-t, \beta) & (t < 0) \end{aligned} \qquad (3.23)$$

where α and β have independent uniform distributions over $(0, 1)$, we shall obtain a distribution of $\xi(t, \alpha, \beta)$ where t runs over the whole real infinite line. There is a well-known mathematical device to map a square on a line segment in such a way that area goes into length. All we need to do is to write our coordinates in the square in the decimal form:

$$\left. \begin{array}{l} \alpha = .\alpha_1\alpha_2 \cdots \alpha_n \cdots \\ \beta = .\beta_1\beta_2 \cdots \beta_n \cdots \end{array} \right\} \tag{3.24}$$

and to put

$$\gamma = .\alpha_1\beta_1\alpha_2\beta_2 \cdots \alpha_n\beta_n \cdots$$

and we obtain a mapping of this sort which is one-one for almost all points both in the line segment and the square. Using this substitution, we define

$$\xi(t, \gamma) = \xi(t, \alpha, \beta) \tag{3.25}$$

We now wish to define

$$\int_{-\infty}^{\infty} K(t) \, d\xi(t, \gamma) \tag{3.26}$$

The obvious thing would be to define this as a Stieltjes [1] integral, but ξ is a very irregular function of t and does not make such a definition possible. If, however, K runs sufficiently rapidly to 0 at $\pm \infty$ and is a sufficiently smooth function, it is reasonable to put

$$\int_{-\infty}^{\infty} K(t) \, d\xi(t, \gamma) = -\int_{-\infty}^{\infty} K'(t)\xi(t, \gamma) \, dt \tag{3.27}$$

Under these circumstances, we have formally

$$\int_0^1 d\gamma \int_{-\infty}^{\infty} K_1(t) \, d\xi(t, \gamma) \int_{-\infty}^{\infty} K_2(t) \, d\xi(t, \gamma)$$

$$= \int_0^1 d\gamma \int_{-\infty}^{\infty} K_1'(t)\xi(t, \gamma) \, dt \int_{-\infty}^{\infty} K_2'(t)\xi(t, \gamma) \, dt$$

$$= \int_{-\infty}^{\infty} K_1'(s) \, ds \int_{-\infty}^{\infty} K_2'(t) \, dt \int_0^1 \xi(s, \gamma)\xi(t, \gamma) \, d\gamma \tag{3.28}$$

Now, if s and t are of opposite signs,

$$\int_0^1 \xi(s, \gamma)\xi(t, \gamma) \, d\gamma = 0 \tag{3.29}$$

---

[1] Stieltjes, T. J. *Annales de la Fac. des Sc. de Toulouse*, 1894, p. 165; Lebesgue, H., *Leçons sur l'Intégration*, Gauthier-Villars et Cie, Paris, 1928.

while if they are of the same sign, and $|s| < |t|$,

$$\int_0^1 \xi(s,\gamma)\xi(t,\gamma)\,d\gamma = \int_0^1 x(|s|,\alpha)x(|t|,\alpha)\,d\alpha$$

$$= \frac{1}{2\pi\sqrt{|s|(|t|-|s|)}} \int_{-\infty}^{\infty} du \int_{-\infty}^{\infty} dv\, uv \exp\left[-\frac{u^2}{2|s|} - \frac{(v-u)^2}{2(|t|-|s|)}\right]$$

$$= \frac{1}{\sqrt{2\pi|s|}} \int_{-\infty}^{\infty} u^2 \exp\left(-\frac{u^2}{2|s|}\right) du$$

$$= |s|\frac{1}{\sqrt{2\pi}} \int_{-\infty}^{\infty} u^2 \exp\left(-\frac{u^2}{2}\right) du = |s| \qquad (3.30)$$

Thus:

$$\int_0^1 d\gamma \int_{-\infty}^{\infty} K_1(t)\,d\xi(t,\gamma) \int_{-\infty}^{\infty} K_2(t)\,d\xi(t,\gamma)$$

$$= -\int_0^{\infty} K_1'(s)\,ds \int_0^s tK_2'(t)\,dt - \int_0^{\infty} K_2'(s)\,ds \int_0^s tK_1'(t)\,dt$$

$$\quad + \int_{-\infty}^0 K_1'(s)\,ds \int_s^0 tK_2'(t)\,dt + \int_{-\infty}^0 K_2'(s)\,ds \int_s^0 tK_1'(t)\,dt$$

$$= -\int_0^{\infty} K_1'(s)\,ds \left[sK_2(s) - \int_0^s K_2(t)\,dt\right]$$

$$\quad - \int_0^{\infty} K_2'(s)\,ds \left[sK_1(s) - \int_0^s K_1(t)\,dt\right]$$

$$\quad + \int_{-\infty}^0 K_1'(s)\,ds \left[-sK_2(s) - \int_s^0 K_2(t)\,dt\right]$$

$$\quad + \int_{-\infty}^0 K_2'(s)\,ds \left[-sK_1(s) - \int_s^0 K_1(t)\,dt\right]$$

$$= -\int_{-\infty}^{\infty} s\,d\,[K_1(s)K_2(s)] = \int_{-\infty}^{\infty} K_1(s)K_2(s)\,ds \qquad (3.31)$$

In particular,

$$\int_0^1 d\gamma \int_{-\infty}^{\infty} K(t+\tau_1)\,d\xi(t,\gamma) \int_{-\infty}^{\infty} K(t+\tau_2)\,d\xi(t,\gamma)$$

$$= \int_{-\infty}^{\infty} K(s)K(s+\tau_2-\tau_1)\,ds \qquad (3.32)$$

Moreover,

$$\int_0^1 d\gamma \prod_{k=1}^n \int_{-\infty}^\infty K(t + \tau_k) \, d\xi(t, \gamma)$$

$$= \sum \prod \int_{-\infty}^\infty K(s)K(s + \tau_j - \tau_k) \, ds \qquad (3.33)$$

where the sum is over all partitions of $\tau_l, \cdots, \tau_n$ into pairs, and the product is over the pairs in each partition.

The expression

$$\int_{-\infty}^\infty K(t + \tau) \, d\xi(\tau, \gamma) = f(t, \gamma) \qquad (3.34)$$

represents a very important ensemble of time series in the variable t, depending on a parameter of distribution γ. We have just shown what amounts to the statement that all the moments and hence all the statistical parameters of this distribution depend on the function

$$\Phi(\tau) = \int_{-\infty}^\infty K(s)K(s + \tau) \, ds$$

$$= \int_{-\infty}^\infty K(s + t)K(s + t + \tau) \, ds \qquad (3.35)$$

which is the statisticians' autocorrelation function with lag τ. Thus the statistics of distribution of $f(t, \gamma)$ are the same as the statistics of $f(t + t_1, \gamma)$; and it can be shown in fact that if

$$f(t + t_1, \gamma) = f(t, \Gamma) \qquad (3.36)$$

then the transformation of γ into Γ preserves measure. In other words, our time series $f(t, \gamma)$ is in statistical equilibrium.

Moreover, if we consider the average of

$$\left[\int_{-\infty}^\infty K(t - \tau) \, d\xi(t, \gamma) \right]^m \left[\int_{-\infty}^\infty K(t + \sigma - \tau) \, d\xi(t, \gamma) \right]^n \qquad (3.37)$$

it will consist of precisely the terms in

$$\int_0^1 d\gamma \left[\int_{-\infty}^\infty K(t - \tau) \, d\xi(t, \gamma) \right]^m \int_0^1 d\gamma \left[\int_{-\infty}^\infty K(t + \sigma - \tau) \, d\xi(t, \gamma) \right]^n$$
$$(3.38)$$

together with a finite number of terms involving as factors powers of

$$\int_{-\infty}^\infty K(\sigma + \tau)K(\tau) \, d\tau \qquad (3.39)$$

and if this approaches 0 when $\sigma \to \infty$, Expression 3.38 will be the limit of Expression 3.37 under these circumstances. In other words, $f(t, \gamma)$ and $f(t + \sigma, \gamma)$ are asymptotically independent in their distributions as $\sigma \to \infty$. By a more generally phrased but entirely similar argument, it may be shown that the simultaneous distribution of $f(t_1, \gamma), \cdots, f(t_n, \gamma)$ and of $f(\sigma + s_1, \gamma), \cdots, f(\sigma + s_m, \gamma)$ tends to the joint distribution of the first and the second set as $\sigma \to \infty$. In other words, any bounded measurable functional or quantity depending on the entire distribution of the values of the function of t, $f(t, \gamma)$, which we may write in the form $\mathscr{F}[f(t, \gamma)]$, must have the property that

$$\lim_{\sigma \to \infty} \int_0^1 \mathscr{F}[f(t, \gamma)]\mathscr{F}[f(t + \sigma, \gamma)] \, d\gamma = \left\{ \int_0^1 \mathscr{F}[f(t, \gamma)] \, d\gamma \right\}^2 \quad (3.40)$$

If now $\mathscr{F}[f(t, \gamma)]$ is invariant under a translation of t, and only takes on the values 0 or 1, we shall have

$$\int_0^1 \mathscr{F}[f(t, \gamma)] \, d\gamma = \int_0^1 \left\{ \mathscr{F}[f(t, \gamma)] \, d\gamma \right\}^2 \quad (3.41)$$

so that the transformation group of $f(t, \gamma)$ into $f(t + \sigma, \gamma)$ is *metrically transitive*. It follows that if $\mathscr{F}[f(t, \gamma)]$ is any integrable functional of f as a function of t, then by the ergodic theorem

$$\int_0^1 \mathscr{F}[f(t, \gamma)] \, d\gamma = \lim_{T \to \infty} \frac{1}{T} \int_0^T \mathscr{F}[f(t, \gamma)] \, dt$$

$$= \lim_{T \to \infty} \frac{1}{T} \int_{-T}^0 \mathscr{F}[f(t, \gamma)] \, dt \quad (3.42)$$

for all values of γ except for a set of zero measure. That is, we can almost always read off any statistical parameter of such a time series, and indeed any denumerable set of statistical parameters, from the past history of a single example. Actually, for such a time series, when we know

$$\lim_{T \to \infty} \frac{1}{T} \int_{-T}^0 f(t, \gamma) f(t - \tau, \gamma) \, dt \quad (3.43)$$

we know $\Phi(t)$ in almost every case, and we have a complete statistical knowledge of the time series.

There are certain quantities dependent on a time series of this sort which have quite interesting properties. In particular, it is interesting to know the average of

$$\exp \left[i \int_{-\infty}^{\infty} K(t) \, d\xi(t, \gamma) \right] \quad (3.44)$$

Formally, this may be written

$$\int_0^1 d\gamma \sum_{n=0}^{\infty} \frac{i^n}{n!} \left[\int_{-\infty}^{\infty} K(t) \, d\xi(t, \gamma) \right]^n$$

$$= \sum_m \frac{(-1)^m}{(2m)!} \left\{ \int_{-\infty}^{\infty} [K(t)]^2 \, dt \right\}^m (2m-1)(2m-3)\cdots 5\cdot 3\cdot 1$$

$$= \sum_m^{\infty} \frac{(-1)^m}{2^m m!} \left\{ \int_{-\infty}^{\infty} [K(t)]^2 \, dt \right\}^m$$

$$= \exp \left\{ -\frac{1}{2} \int_{-\infty}^{\infty} [K(t)]^2 \, dt \right\} \tag{3.45}$$

It is a very interesting problem to try to build up a time series as general as possible from the simple Brownian motion series. In such constructions, the example of the Fourier developments suggests that such expansions as Expression 3.44 are convenient building blocks for this purpose. In particular, let us investigate time series of the special form

$$\int_a^b d\lambda \exp \left[i \int_{-\infty}^{\infty} K(t+\tau, \lambda) \, d\xi(\tau, \gamma) \right] \tag{3.46}$$

Let us suppose that we know $\xi(\tau, \gamma)$ as well as Expression 3.46. Then, as in Eq. 3.45, if $t_1 > t_2$,

$$\int_0^1 d\gamma \exp \left\{ is[\xi(t_1, \gamma) - \xi(t_2, \gamma)] \right\}$$

$$\times \int_a^b d\lambda \exp \left[i \int_{-\infty}^{\infty} K(t+\tau, \lambda) \, d\xi(t, \gamma) \right]$$

$$= \int_a^b d\lambda \exp \left\{ -\frac{1}{2} \int_{-\infty}^{\infty} [K(t+\tau, \lambda)]^2 \, dt \right.$$

$$\left. - \frac{s^2}{2}(t_2 - t_1) - s \int_{t_2}^{t_1} K(t, \lambda) \, dt \right\} \tag{3.47}$$

If we now multiply by $\exp[s^2(t_2 - t_1)/2]$, let $s(t_2 - t_1) = i\sigma$, and let $t_2 \to t_1$, we obtain

$$\int_a^b d\lambda \exp \left\{ -\frac{1}{2} \int_{-\infty}^{\infty} [K(t+\tau, \lambda)]^2 \, dt - i\sigma K(t_1, \lambda) \right\} \tag{3.48}$$

Let us take $K(t_1, \lambda)$ and a new independent variable μ and solve for λ, obtaining

$$\lambda = Q(t_1, \mu) \tag{3.49}$$

Then Expression 3.48 becomes

$$\int_{K(t_1,a)}^{K(t_1,b)} e^{i\mu\sigma} \, d\mu \, \frac{\partial Q(t_1, \mu)}{\partial \mu} \exp\left(-\frac{1}{2} \int_{-\infty}^{\infty} \{K[t + \tau, Q(t_1, \mu)]\}^2 \, dt\right) \quad (3.50)$$

From this, by a Fourier transformation, we can determine

$$\frac{\partial Q(t_1, \mu)}{\partial \mu} \exp\left(-\frac{1}{2} \int_{-\infty}^{\infty} \{K[t + \tau, Q(t_1, \mu)]\}^2 \, dt\right) \quad (3.51)$$

as a function of μ when μ lies between $K(t_1, a)$ and $K(t_1, b)$. If we integrate this function with respect to μ, we determine

$$\int_a^\lambda d\lambda \exp\left\{-\frac{1}{2} \int_{-\infty}^{\infty} [K(t + \tau, \lambda)]^2 \, dt\right\} \quad (3.52)$$

as a function of $K(t_1, \lambda)$ and t_1. That is, there is a known function $F(u, v)$, such that

$$\int_a^\lambda d\lambda \exp\left\{-\frac{1}{2} \int_{-\infty}^{\infty} [K(t + \tau, \lambda)]^2 \, dt\right\} = F[K(t_1, \lambda), t_1] \quad (3.53)$$

Since the left-hand side of this equation does not depend on t_1, we may write it $G(\lambda)$, and put

$$F[K(t_1, \lambda), t_1] = G(\lambda) \quad (3.54)$$

Here, F is a known function, and we can invert it with respect to the first argument, and put

$$K(t_1, \lambda) = H[G(\lambda), t_1] \quad (3.55)$$

where it is also a known function. Then

$$G(\lambda) = \int_a^\lambda d\lambda \exp\left(-\frac{1}{2} \int_{-\infty}^{\infty} \{H[G(\lambda), t + \tau]\}^2 \, dt\right) \quad (3.56)$$

Then the function

$$\exp\left\{-\frac{1}{2} \int_{-\infty}^{\infty} [H(u, t)]^2 \, dt\right\} = R(u) \quad (3.57)$$

will be a known function, and

$$\frac{dG}{d\lambda} = R(G) \quad (3.58)$$

That is,

$$\frac{dG}{R(G)} = d\lambda \quad (3.59)$$

or

$$\lambda = \int \frac{dG}{R(G)} + \text{const} = S(G) + \text{const} \tag{3.60}$$

This constant will be given by

$$G(a) = 0 \tag{3.61}$$

or

$$a = S(0) + \text{const} \tag{3.62}$$

It is easy to see that if a is finite, it does not matter what value we give it; for our operator is not changed if we add a constant to all values of λ. We can hence make it 0. We have thus determined λ as a function of G, and thus G as a function of λ. Thus, by Eq. 3.55, we have determined $K(t, \lambda)$. To finish the determination of Expression 3.46, we need only know b. This can be determined, however, by a comparison of

$$\int_a^b d\lambda \exp \left\{ -\frac{1}{2} \int_{-\infty}^{\infty} [K(t, \lambda)]^2 \, dt \right\} \tag{3.63}$$

with

$$\int_0^1 d\gamma \int_a^b d\lambda \exp \left[i \int_{-\infty}^{\infty} K(t, \lambda) \, d\xi(t, \gamma) \right] \tag{3.64}$$

Thus, under certain circumstances which remain to be definitely formulated, if a time series may be written in the form of Expression 3.46 and we know $\xi(t, \gamma)$ as well, we can determine the function $K(t, \lambda)$ in Expression 3.46 and the numbers a and b, except for an undetermined constant added to a, λ, and b. There is no extra difficulty if $b = +\infty$, and it is not hard to extend the reasoning to the case where $a = -\infty$. Of course, a good deal of work remains to be done to discuss the problem of the inversion of the functions inverted when the results are not single-valued, and the general conditions of validity of the expansions concerned. Still, we have at least taken a first step toward the solution of the problem of reducing a large class of time series to a canonical form, and this is most important for the concrete formal application of the theories of prediction and of the measurement of information, as we have sketched them earlier in this chapter.

There is still one obvious limitation which we should remove from this approach to the theory of time series: the necessity which we are under of knowing $\xi(t, \gamma)$ as well as the time series which we are expanding in the form of Expression 3.46. The question is: under what circumstances can we represent a time series of known statistical parameters as determined by a Brownian motion; or at least as the limit in some sense or other of time series determined by Brownian motions? We shall confine ourselves to time series with the property of metrical transitivity, and with the even stronger property that if we take intervals of fixed length but remote in time, the distributions of any functionals of the segments of the time series in these intervals approach independence as the intervals recede from each other.[1] The theory to be developed here has already been sketched by the author.

If $K(t)$ is a sufficiently continuous function, it is possible to show that the zeros of

$$\int_{-\infty}^{\infty} K(t + \tau) \, d\xi(\tau, \gamma) \tag{3.65}$$

almost always have a definite density, by a theorem of M. Kac, and that this density can be made as great as we wish by a proper choice of K. Let K_D be so selected that this density is D. Then the sequence of zeros of $\int_{-\infty}^{\infty} K_D(t + \tau) \, d\xi(\tau, \gamma)$ from $-\infty$ to ∞ will be called $Z_n(D, \gamma)$, $-\infty < n < \infty$. Of course, in the numeration of these zeros, n is determined except for an additive constant integer.

Now, let $T(t, \mu)$ be any time series in the continuous variable t, while μ is a parameter of distribution of the time series, varying uniformly over $(0, 1)$. Then let

$$T_D(t, \mu, \gamma) = T[t - Z_n(D, \gamma), \mu] \tag{3.66}$$

where the Z_n taken is the one just preceding t. It will be seen that for any finite set of values $t_1, t_2, \cdots, t_v$ of x the simultaneous distribution of $T_D(t_\kappa, \mu, \gamma)$ ($\kappa = 1, 2, \cdots, v$) will approach the simultaneous distribution of $T(t_\kappa, \mu)$ for the same t_κ's as $D \to \infty$, for almost every value of μ. However, $T_D(t, \mu, \gamma)$ is completely determined by t, μ, D, and $\xi(\tau, \gamma)$. It is therefore not inappropriate to try to express $T_D(t, \mu, \gamma)$, for a given D and a given μ, either directly in the form of Expression 3.46 or in some way or another as a time series which has a distribution which is a limit (in the loose sense just given) of distributions of this form.

It must be admitted that this is a program to be carried through in the future, rather than one which we can consider as already

[1] This is the mixing property of Koopman, which is the necessary and sufficient ergodic assumption to justify statistical mechanics.

accomplished. Nevertheless, it is the program which, in the opinion of the author, offers the best hope for a rational, consistent treatment of the many problems associated with non-linear prediction, non-linear filtering, the evaluation of the transmission of information in non-linear situations, and the theory of the dense gas and turbulence. Among these problems are perhaps the most pressing facing communication engineering.

Let us now come to the prediction problem for time series of the form of Eq. 3.34. We see that the only independent statistical parameter of the time series is $\Phi(t)$, as given by Eq. 3.35; which means that the only significant quantity connected with $K(t)$ is

$$\int_{-\infty}^{\infty} K(s)K(s + t)\, ds \tag{3.67}$$

Here of course K is real.

Let us put

$$K(s) = \int_{-\infty}^{\infty} k(\omega)e^{\iota \omega s}\, d\omega \tag{3.68}$$

employing a Fourier transformation. To know $K(s)$ is to know $k(\omega)$, and vice versa. Then

$$\frac{1}{2\pi} \int_{-\infty}^{\infty} K(s)K(s + \tau)\, ds = \int_{-\infty}^{\infty} k(\omega)k(-\omega)e^{\iota \omega \tau}\, d\omega \tag{3.69}$$

Thus a knowledge of $\Phi(\tau)$ is tantamount to a knowledge of $k(\omega)k(-\omega)$. Since, however, $K(s)$ is real,

$$K(s) = \int_{-\infty}^{\infty} \overline{k(\omega)}e^{-\iota \omega s}\, d\omega \tag{3.70}$$

whence $k(\omega) = k(-\omega)$. Thus $|k(\omega)|^2$ is a known function, which means that the real part of $\log|k(\omega)|$ is a known function.

If we write

$$F(\omega) = \mathscr{R}\{\log[k(\omega)]\} \tag{3.71}$$

then the determination of $K(s)$ is equivalent to the determination of the imaginary part of $\log k(\omega)$. This problem is not determinate unless we put some further restriction on $k(\omega)$. The type of restriction which we shall put is that $\log k(s)$ shall be analytic and of a sufficiently small rate of growth for ω in the upper half-plane. In order to make this restriction, $k(\omega)$ and $[k(\omega)]^{-1}$ will be assumed to be of algebraic growth on the real axis. Then $[F(\omega)]^2$ will be even

and at most logarithmically infinite, and the Cauchy principal value
of

$$G(\omega) = \frac{1}{\pi} \int_{-\infty}^{\infty} \frac{F(u)}{u - \omega} \, du \qquad (3.72)$$

will exist. The transformation indicated by Eq. 3.72, known as the
Hilbert transformation, changes $\cos \lambda\omega$ into $\sin \lambda\omega$ and $\sin \lambda\omega$ into
$-\cos \lambda\omega$. Thus $F(\omega) + iG(\omega)$ is a function of the form

$$\int_{0}^{\infty} e^{i\lambda\omega} \, d[M(\lambda)] \qquad (3.73)$$

and satisfies the required conditions for $\log |k(\omega)|$ in the lower
half-plane. If we now put

$$k(\omega) = \exp\left[F(\omega) + iG(\omega)\right] \qquad (3.74)$$

it can be shown that $k(\omega)$ is a function which, under very general
conditions, is such that $K(s)$, as defined in Eq. 3.68, vanishes for all
negative arguments. Thus

$$f(t, \gamma) = \int_{-t}^{\infty} K(t + \tau) \, d\xi(\tau, \gamma) \qquad (3.75)$$

On the other hand, it can be shown that we may write $1/k(\omega)$ in the
form

$$\lim_{n \to \infty} \int_{0}^{\infty} e^{i\lambda\omega} \, dN_n(\lambda) \qquad (3.76)$$

where the N_n's are properly determined; and that this can be done
in such a way that

$$\xi(\tau, \gamma) = \lim_{n \to \infty} \int_{0}^{\tau} dt \int_{-t}^{\infty} Q_n(t + \sigma) f(\sigma, \gamma) \, d\sigma \qquad (3.77)$$

Here the Q_n's must have the formal property that

$$f(t, \gamma) = \lim_{n \to \infty} \int_{-t}^{\infty} K(t + \tau) \, d\tau \int_{-\tau}^{\infty} Q_n(\tau + \sigma) f(\sigma, \gamma) \, d\sigma \qquad (3.78)$$

In general, we shall have

$$\psi(t) = \lim_{n \to \infty} \int_{-t}^{\infty} K(t + \tau) \, d\tau \int_{-\tau}^{\infty} Q_n(\tau + \sigma) \psi(\sigma) \, d\sigma \qquad (3.79)$$

or if we write (as in Eq. 3.68)

$$K(s) = \int_{-\infty}^{\infty} k(\omega)e^{i\omega s}\, d\omega$$

$$Q_n(s) = \int_{-\infty}^{\infty} q_n(\omega)e^{i\omega s}\, d\omega$$

$$\psi(s) = \int_{-\infty}^{\infty} \Psi(\omega)e^{i\omega s}\, d\omega \qquad (3.80)$$

then

$$\Psi(\omega) = \lim_{n\to\infty} (2\pi)^{3/2}\Psi(\omega)q_n(-\omega)k(\omega) \qquad (3.81)$$

Thus

$$\lim_{n\to\infty} q_n(-\omega) = \frac{1}{(2\pi)^{3/2}k(\omega)} \qquad (3.82)$$

We shall find this result useful in getting the operator of prediction into a form concerning frequency rather than time.

Thus the past and present of $\xi(t, \gamma)$, or properly of the "differential" $d\xi(t, \gamma)$, determine the past and present of $f(t, \gamma)$, and vice versa.

Now, if $A > 0$,

$$f(t + A, \gamma) = \int_{-t-A}^{\infty} K(t + A + \tau)\, d\xi(\tau, \gamma)$$

$$= \int_{-t-A}^{-t} K(t + A + \tau)\, d\xi(\tau, \gamma)$$

$$+ \int_{-t}^{\infty} K(t + A + \tau)\, d\xi(\tau, \gamma) \qquad (3.83)$$

Here the first term of the last expression depends on a range of $d\xi(\tau, \gamma)$ of which a knowledge of $f(\sigma, \gamma)$ for $\sigma \leqslant t$ tells us nothing, and is entirely independent of the second term.   Its mean square value is

$$\int_{-t-A}^{t} [K(t + A + \tau)]^2\, d\tau = \int_0^A [K(\tau)]^2\, d\tau \qquad (3.84)$$

and this tells us all there is to know about it statistically.   It may be shown to have a Gaussian distribution with this mean square value.   It is the error of the best possible prediction of $f(t + A, \gamma)$.

The best possible prediction itself is the last term of Eq. 3.83,

$$\int_{-t}^{\infty} K(t + A + \tau)\, d\xi(\tau, \gamma)$$

$$= \lim_{n\to\infty} \int_{-t}^{\infty} K(t + A + \tau)\, d\tau \int_{-\tau}^{\infty} Q_n(\tau + \sigma)f(\sigma, \gamma)\, d\sigma \qquad (3.85)$$

If we now put

$$k_A(\omega) = \frac{1}{2\pi} \int_0^\infty K(t + A)e^{-i\omega t}\, dt \qquad (3.86)$$

and if we apply the operator of Eq. 3.85 to $e^{i\omega t}$, obtaining

$$\lim_{n \to \infty} \int_{-t}^\infty K(t + A + \tau)\, d\tau \int_{-\tau}^\infty Q_n(\tau + \sigma)e^{i\omega\sigma}\, d\sigma = A(\omega)e^{i\omega t} \qquad (3.87)$$

we shall find out (somewhat as in Eq. 3.81) that

$$A(\omega) = \lim_{n \to \infty} (2\pi)^{3/2}\, q_n(-\omega)k_A(\omega)$$

$$= k_A(\omega)/k(\omega)$$

$$= \frac{1}{2\pi k(\omega)} \int_A^\infty e^{-i\omega(t-A)}\, dt \int_{-\infty}^\infty k(u)e^{iut}\, du \qquad (3.88)$$

This is then the frequency form of the best prediction operator.

The problem of filtering in the case of time series such as Eq. 3.34 is very closely allied to the prediction problem. Let our message plus noise be of the form

$$m(t) + n(t) = \int_0^\infty K(\tau)\, d\xi(t - \tau, \gamma) \qquad (3.89)$$

and let the message be of the form

$$m(t) = \int_{-\infty}^\infty Q(\tau)\, d\xi(t - \tau, \gamma) + \int_{-\infty}^\infty R(\tau)\, d\xi(t - \tau, \delta) \qquad (3.90)$$

where γ and δ are distributed independently over $(0, 1)$. Then the predictable part of the $m(t + a)$ is clearly

$$\int_0^\infty Q(\tau + a)\, d\xi(t - \tau, \gamma) \qquad (3.901)$$

and the mean square error of prediction is

$$\int_{-\infty}^a [Q(\tau)]^2\, d\tau + \int_{-\infty}^\infty [R(\tau)]^2\, d\tau \qquad (3.902)$$

Furthermore, let us suppose that we know the following quantities:

$$\phi_{22}(t) = \int_0^1 dy \int_0^1 d\delta \, n(t + \tau)n(\tau)$$

$$= \int_{-\infty}^{\infty} [K(|t| + \tau) - Q(|t| + \tau)][K(\tau) - Q(\tau)] \, d\tau$$

$$= \int_0^{\infty} [K(|t| + \tau) - Q(|t| + \tau)][K(\tau) - Q(\tau)] \, d\tau$$

$$+ \int_{-|t|}^0 [K(|t| + \tau) - Q(|t| + \tau)][-Q(\tau)] \, d\tau$$

$$+ \int_{-\infty}^{-|t|} Q(|t| + \tau)Q(\tau) \, d\tau + \int_{-\infty}^{\infty} R(|t| + \tau)R(\tau) \, d\tau$$

$$= \int_0^{\infty} K(|t| + \tau)K(\tau) \, d\tau - \int_{-|t|}^{\infty} K(|t| + \tau)Q(\tau) \, d\tau$$

$$+ \int_{-\infty}^{\infty} Q(|t| + \tau)Q(\tau) \, d\tau + \int_{-\infty}^{\infty} R(|t| + \tau)R(\tau) \, d\tau$$

$$\tag{3.903}$$

$$\phi_{11}(\tau) = \int_0^1 dy \int_0^1 d\delta \, m(|t| + \tau)m(\tau)$$

$$= \int_{-\infty}^{\infty} Q(|t| + \tau)Q(\tau) \, d\tau + \int_{-\infty}^{\infty} R(|t| + \tau)R(\tau) \, d\tau \tag{3.904}$$

$$\phi_{12}(\tau) = \int_0^1 dy \int_0^1 d\delta \, m(t + \tau)n(\tau)$$

$$= \int_0^1 dy \int_0^1 d\delta \, m(t + \tau)[m(\tau) + n(\tau)] - \phi_{11}(\tau)$$

$$= \int_0^1 dy \int_{-t}^{\infty} K(\sigma + t) \, d\xi(\tau - \sigma, \gamma) \int_{-t}^{\infty} Q(\tau) \, d\xi(\tau - \sigma, \gamma) - \phi_{11}(\tau)$$

$$= \int_{-t}^{\infty} K(t + \tau)Q(\tau) \, d\tau - \phi_{11}(\tau) \tag{3.905}$$

The Fourier transforms of these three quantities are, respectively,

$$\left. \begin{aligned} \Phi_{22}(\omega) &= |k(\omega)|^2 + |q(\omega)|^2 - q(\omega)\overline{k(\omega)} - k(\omega)\overline{q(\omega)} + |r(\omega)|^2 \\ \Phi_{11}(\omega) &= |q(\omega)|^2 + |r(\omega)|^2 \\ \Phi_{12}(\omega) &= k(\omega)\overline{q(\omega)} - |q(\omega)|^2 - |r(\omega)| \end{aligned} \right\} \tag{3.906}$$

where

$$
\left.\begin{aligned}
k(\omega) &= \frac{1}{2\pi} \int_0^\infty K(s)e^{-\iota\omega s}\, ds \\[4pt]
q(\omega) &= \frac{1}{2\pi} \int_{-\infty}^\infty \overline{Q(s)}e^{-\iota\omega s}\, ds \\[4pt]
r(\omega) &= \frac{1}{2\pi} \int_{-\infty}^\infty R(s)e^{-\iota\omega s}\, ds
\end{aligned}\right\} \tag{3.907}
$$

That is,

$$
\Phi_{11}(\omega) + \Phi_{12}(\omega) + \overline{\Phi_{12}(\omega)} + \overline{\Phi_{22}(\omega)} = |k(\omega)|^2 \tag{3.908}
$$

and

$$
q(\omega)\overline{k(\omega)} = \Phi_{11}(\omega) + \Phi_{21}(\omega) \tag{3.909}
$$

where for symmetry we write $\Phi_{21}(\omega) = \overline{\Phi_{12}(\omega)}$. We can now determine $k(\omega)$ from Eq. 3.908, as we have defined $k(\omega)$ before on the basis of Eq. 3.74. Here we put $\Phi(t)$ for $\Phi_{11}(t) + \Phi_{22}(t) + 2\mathscr{R}[\Phi_{12}(t)]$. This will give us

$$
q(\omega) = \frac{\Phi_{11}(\omega) + \Phi_{21}(\omega)}{\overline{k(\omega)}} \tag{3.910}
$$

Hence

$$
Q(t) = \int_{-\infty}^\infty \frac{\Phi_{11}(\omega) + \Phi_{21}(\omega)}{\overline{k(\omega)}} e^{\iota\omega t}\, d\omega \tag{3.911}
$$

and thus the best determination of $m(t)$, with the least mean square error, is

$$
\int_0^\infty d\xi(t - \tau, \gamma) \int_{-\infty}^\infty \frac{\Phi_{11}(\omega) + \Phi_{21}(\omega)}{\overline{k(\omega)}} e^{\iota\omega(t+a)}\, d\omega \tag{3.912}
$$

Combining this with Eq. 3.89, and using an argument similar to the one by which we obtained Eq. 3.88, we see that the operator on $m(t) + n(t)$ by which we obtain the "best" representation of $m(t + a)$, if we write it on the frequency scale, is

$$
\frac{1}{2\pi k(\omega)} \int_a^\infty e^{-\iota\omega(t-a)}\, dt \int_{-\infty}^\infty \frac{\Phi_{11}(u) + \Phi_{21}(u)}{\overline{k(u)}} e^{\iota u t}\, du \tag{3.913}
$$

This operator constitutes a characteristic operator of what electrical engineers know as a *wave filter*. The quantity a is the *lag* of the filter. It may be either positive or negative; when it is negative, $-a$ is known as the *lead*. The apparatus corresponding to Expression 3.913 may always be constructed with as much accuracy as we like. The details of its construction are more for the specialist in electrical engineering than for the reader of this book. They may be found elsewhere.[1]

The mean square filtering error (Expression 3.902) may be represented as the sum of the mean square filtering error for infinite lag:

$$\int_{-\infty}^{\infty} [R(\tau)]^2 \, d\tau = \Phi_{11}(0) - \int_{-\infty}^{\infty} [Q(\tau)]^2 \, d\tau$$

$$= \frac{1}{2\pi} \int_{-\infty}^{\infty} \Phi_{11}(\omega) \, d\omega - \frac{1}{2\pi} \int_{-\infty}^{\infty} \left| \frac{\Phi_{11}(\omega) + \Phi_{21}(\omega)}{\overline{k}(\omega)} \right|^2 d\omega$$

$$= \frac{1}{2\pi} \int_{-\infty}^{\infty} \left[\Phi_{11}(\omega) - \frac{|\Phi_{11}(\omega) + \Phi_{21}(\omega)|^2}{\Phi_{11}(\omega) + \Phi_{12}(\omega) + \Phi_{21}(\omega) + \Phi_{22}(\omega)} \right] d\omega$$

$$= \frac{1}{2\pi} \int_{-\infty}^{\infty} \frac{\begin{vmatrix} \Phi_{11}(\omega) & \Phi_{12}(\omega) \\ \Phi_{21}(\omega) & \Phi_{22}(\omega) \end{vmatrix}}{\Phi_{11}(\omega) + \Phi_{12}(\omega) + \Phi_{21}(\omega) + \Phi_{22}(\omega)} \, d\omega \qquad (3.914)$$

and a part dependent on the lag:

$$\int_{-\infty}^{a} [Q(\tau)]^2 \, dt = \int_{-\infty}^{a} dt \left| \int_{-\infty}^{\infty} \frac{\Phi_{11}(\omega) + \Phi_{21}(\omega)}{\overline{k}(\omega)} e^{i\omega t} \, d\omega \right|^2 \qquad (3.915)$$

It will be seen that the mean square error of filtering is a monotonely decreasing function of lag.

Another question which is interesting in the case of messages and noises derived from the Brownian motion is the matter of rate of transmission of information. Let us consider for simplicity the case where the message and the noise are incoherent, that is, when

$$\Phi_{12}(\omega) \equiv \Phi_{21}(\omega) \equiv 0 \qquad (3.916)$$

In this case, let us consider

$$\left. \begin{aligned} m(t) &= \int_{-\infty}^{\infty} M(\tau) \, d\xi(t - \tau, \gamma) \\ n(t) &= \int_{-\infty}^{\infty} N(\tau) \, d\xi(t - \tau, \delta) \end{aligned} \right\} \qquad (3.917)$$

where γ and δ are distributed independently. Let us suppose we know $m(t) + n(t)$ over $(-A, A)$; how much information do we have concerning $m(t)$? Note that we should heuristically expect that it

---

[1] We refer especially to recent papers by Dr. Y. W. Lee.

would not be very different from the amount of information concerning

$$\int_{-A}^{A} M(\tau) \, d\xi(t - \tau, \gamma) \qquad (3.918)$$

which we have when we know all values of

$$\int_{-A}^{A} M(\tau) \, d\xi(t - \tau, \gamma) + \int_{-A}^{A} N(\tau) \, d\xi(t - \tau, \delta) \qquad (3.919)$$

where γ and δ have independent distributions. It can, however, be shown that the nth Fourier coefficient of Expression 3.918 has a Gaussian distribution independent of all the other Fourier coefficients, and that its mean square value is proportional to

$$\left| \int_{-A}^{A} M(\tau) \exp\left(i \frac{\pi n \tau}{A} \right) d\tau \right|^2 \qquad (3.920)$$

Thus, by Eq. 3.09, the total amount of information available concerning M is

$$\sum_{n=-\infty}^{\infty} \frac{1}{2} \log_2 \frac{\left| \int_{-A}^{A} M(\tau) \exp\left(i \frac{\pi n \tau}{A} \right) d\tau \right|^2 + \left| \int_{-A}^{A} N(\tau) \exp\left(i \frac{\pi n \tau}{A} \right) d\tau \right|^2}{\left| \int_{-A}^{A} N(\tau) \exp\left(i \frac{\pi n \tau}{A} \right) d\tau \right|^2} \qquad (3.921)$$

and the time density of communication of energy is this quantity divided by $2A$. If now $A \to \infty$, Expression 3.921 approaches

$$\frac{1}{2\pi} \int_{-\infty}^{\infty} du \log_2 \frac{\left| \int_{-\infty}^{\infty} M(\tau) \exp iu\tau \, d\tau \right|^2 + \left| \int_{-\infty}^{\infty} N(\tau) \exp iu\tau \, d\tau \right|^2}{\left| \int_{-\infty}^{\infty} N(\tau) \exp iu\tau \, d\tau \right|^2} \qquad (3.922)$$

This is precisely the result which the author and Shannon have already obtained for the rate of transmission of information in this case. As will be seen, it depends not only on the width of the frequency band available for transmitting the message but also on the noise level. As a matter of fact, it has a close relation to the audiograms used to measure the amount of hearing and loss of hearing in a given individual. Here the abscissa is frequency, the ordinate of lower boundary is the logarithm of the intensity of the threshold of audible intensity—what we may call the logarithm of the intensity of the *internal noise* of the receiving system—and the upper boundary, the logarithm of the intensity of the greatest message the system is suited to handle. The area between them, which is a quantity of the dimension of Expression 3.922, is then taken as a measure of the rate of transmission of information with which the ear is competent to cope.

The theory of messages depending linearly on the Brownian motion has many important variants. The key formulae are Eqs. 3.88 and 3.914 and Expression 3.922, together, of course, with the definitions necessary to interpret these. There are a number of variants of this theory. First: the theory gives us the best possible design of predictors and of wave filters in the case in which the messages and the noises represent the response of linear resonators to Brownian motions; but in much more general cases, they represent a possible design for predictors and filters. This will not be an absolute best possible design, but it will minimize the mean square error of prediction and filtering, in so far as this can be done with apparatus performing linear operations. However, there will generally be some non-linear apparatus which gives a performance still better than that of any linear apparatus.

Next, the time series here have been simple time series, in which a single numerical variable depends on the time. There are also multiple time series, in which a number of such variables depend simultaneously on the time; and it is these which are of greatest importance in economics, meteorology, and the like. The complete weather map of the United States, taken from day to day, constitutes such a time series. In this case, we have to develop a number of functions simultaneously in terms of the frequency, and the quadratic quantities such as Eq. 3.35 and the $|k(\omega)|^2$ of the arguments following Eq. 3.70 are replaced by arrays of pairs of quantities—that is, *matrices*. The problem of determining $k(\omega)$ in terms of $|k(\omega)|^2$, in such a way as to satisfy certain auxiliary conditions in the complex plane, becomes much more difficult, especially as the multiplication of matrices is not a permutable operation. However, the problems involved in this multidimensional theory have been solved, at least in part, by Krein and the author.

The multidimensional theory represents a complication of the one already given. There is another closely related theory which is a simplification of it. This is the theory of prediction, filtering, and amount of information in discrete time series. Such a series is a sequence of functions $f_n(\alpha)$ of a parameter α, where n runs over all integer values from $-\infty$ to ∞. The quantity α is as before the parameter of distribution, and may be taken to run uniformly over (0, 1). The time series is said to be *in statistical equilibrium* when the change of n to $n + \nu$ (ν an integer) is equivalent to a measure-preserving transformation into itself of the interval (0, 1) over which α runs.

The theory of discrete time series is simpler in many respects than the theory of the continuous series. It is much easier, for instance, to make them depend on a sequence of independent choices. Each term (in the mixing case) will be representable as a combination of the previous terms with a quantity independent of all previous terms, distributed uniformly over (0, 1), and the sequence of these independent factors may be taken to replace the Brownian motion which is so important in the continuous case.

If $f_n(\alpha)$ is a time series in statistical equilibrium, and it is metrically transitive, its autocorrelation coefficient will be

$$\phi_m = \int_0^1 f_m(\alpha)f_0(\alpha)\,d\alpha \qquad (3.923)$$

and we shall have

$$\phi_m = \lim_{N\to\infty} \frac{1}{N+1} \sum_0^N f_{k+m}(\alpha)f_k(\alpha)$$

$$= \lim_{N\to\infty} \frac{1}{N+1} \sum_0^N f_{-k+m}(\alpha)f_{-k}(\alpha) \qquad (3.924)$$

for almost all α. Let us put

$$\phi_n = \frac{1}{2\pi} \int_{-\pi}^{\pi} \Phi(\omega)e^{in\omega}\,d\omega \qquad (3.925)$$

or

$$\Phi(\omega) = \sum_{-\infty}^{\infty} \phi_n e^{-in\omega} \qquad (3.926)$$

Let

$$\frac{1}{2}\log \Phi(\omega) = \sum_{-\infty}^{\infty} p_n \cos n\omega \qquad (3.927)$$

and let

$$G(\omega) = \frac{p_0}{2} + \sum_1^{\infty} p_n e^{in\omega} \qquad (3.928)$$

Let

$$e^{G(\omega)} = k(\omega) \qquad (3.929)$$

Then under very general conditions, $k(\omega)$ will be the boundary value on the unit circle of a function without zeros or singularities inside the unit circle if ω is the angle. We shall have

$$|k(\omega)|^2 = \Phi(\omega) \tag{3.930}$$

If now we put for the best linear prediction of $f_n(\alpha)$ with a lead of ν

$$\sum_0^\infty f_{n-\nu}(\alpha) W_\nu \tag{3.931}$$

we shall find that

$$\sum_0^\infty W_\mu e^{i\mu\omega} = \frac{1}{2\pi k(\omega)} \sum_{\mu=\nu}^\infty e^{i\omega(\mu-\nu)} \int_{-\pi}^\pi k(u)e^{-i\mu u}\,du \tag{3.932}$$

This is the analogue of Eq. 3.88. Let us note that if we put

$$k_\mu = \frac{1}{2\pi} \int_{-\pi}^\pi k(u)e^{-i\mu u}\,du \tag{3.933}$$

then

$$\sum_0^\infty W_\mu e^{i\mu\omega} = e^{-i\nu\omega} \frac{\displaystyle\sum_\nu^\infty k_\mu e^{i\mu\omega}}{\displaystyle\sum_0^\infty k_\mu e^{i\mu\omega}}$$

$$= e^{-i\nu\omega}\left(1 - \frac{\displaystyle\sum_0^{\nu-1} k_\mu e^{i\mu\omega}}{\displaystyle\sum_0^\infty k_\mu e^{i\mu\omega}}\right) \tag{3.934}$$

It will clearly be the result of the way we have formed $k(\omega)$ that in a very general set of cases we can put

$$\frac{1}{k(\omega)} = \sum_0^\infty q_\mu e^{i\mu\omega} \tag{3.935}$$

Then Eq. 3.934 becomes

$$\sum_0^\infty W_\mu e^{i\mu\omega} = e^{-i\nu\omega}\left(1 - \sum_0^{\nu-1} k_\mu e^{i\mu\omega} \sum_0^\infty q_\lambda e^{i\lambda\omega}\right) \tag{3.936}$$

In particular, if $\nu = 1$,

$$\sum_0^\infty W_\mu e^{i\mu\omega} = e^{-i\omega}\left(1 - k_0 \sum_0^\infty q_\lambda e^{i\lambda\omega}\right) \tag{3.937}$$

or

$$W_\mu = -q_{\lambda+1}k_0 \qquad (3.938)$$

Thus for a prediction one step ahead, the best value for $f_{n+1}(\alpha)$ is

$$-k_0 \sum_0^\infty q_{\lambda+1}f_{n-\lambda}(\alpha) \qquad (3.939)$$

and by a process of step-by-step prediction, we can solve the entire problem of linear prediction for discrete time series. As in the continuous case, this will be the best prediction possible by any method if

$$f_n(\alpha) = \int_{-\infty}^\infty K(n - \tau)\, d\xi(\tau, \alpha) \qquad (3.940)$$

The transfer of the filtering problem from the continuous to the discrete case follows much the same lines of argument. Formula 3.913 for the frequency characteristic of the best filter takes the form

$$\frac{1}{2\pi k(\omega)} \sum_{\nu=a}^\infty e^{-i\omega(\nu-a)} \int_{-\pi}^\pi \frac{[\Phi_{11}(u) + \Phi_{21}(u)]\, e^{iu\nu}\, du}{k(u)} \qquad (3.941)$$

where all the terms receive the same definitions as in the continuous case, except that all integrals on ω or u are from $-\pi$ to π instead of from $-\infty$ to ∞ and all sums on ν are discrete sums instead of integrals on t. The filters for discrete time series are usually not so much physically constructible devices to be used with an electric circuit as mathematical procedures to enable statisticians to obtain the best results with statistically impure data.

Finally, the rate of transfer of information by a discrete time series of the form

$$\int_{-\infty}^\infty M(n - \tau)\, d\xi(t, \gamma) \qquad (3.942)$$

in the presence of a noise

$$\int_{-\infty}^\infty N(n - \tau)\, d\xi(t, \delta) \qquad (3.943)$$

when γ and δ are independent, will be the precise analogue of Expression 3.922, namely,

$$\frac{1}{2\pi} \int_{-\pi}^\pi du \, \log_2 \frac{\left| \int_{-\infty}^\infty M(\tau)e^{iu\tau}\, d\tau \right|^2 + \left| \int_{-\infty}^\infty N(\tau)e^{iu\tau}\, d\tau \right|^2}{\left| \int_{-\infty}^\infty N(\tau)e^{iu\tau}\, d\tau \right|^2} \qquad (3.944)$$

where over $(-\pi, \pi)$,

$$\left|\int_{-\infty}^{\infty} M(\tau)e^{iu\tau}\,d\tau\right|^2 \tag{3.945}$$

represents the power distribution of the message in frequency, and

$$\left|\int_{-\infty}^{\infty} N(\tau)e^{iu\tau}\,d\tau\right|^2 \tag{3.946}$$

that of the noise.

The statistical theories we have here developed involve a full knowledge of the pasts of the time series we observe. In every case, we have to be content with less, as our observation does not run indefinitely into the past. The development of our theory beyond this point, as a practical statistical theory, involves an extension of existing methods of sampling. The author and others have made a beginning in this direction. It involves all the complexities of the use either of Bayes' law, on the one hand, or of those terminological tricks in the theory of likelihood,[1] on the other, which seem to avoid the necessity for the use of Bayes' law but which in reality transfer the responsibility for its use to the working statistician, or the person who ultimately employs his results. Meanwhile, the statistical theorist is quite honestly able to say that he has said nothing which is not perfectly rigorous and unimpeachable.

Finally, this chapter should end with a discussion of modern quantum mechanics. These represent the highest point of the invasion of modern physics by the theory of time series. In the Newtonian physics, the sequence of physical phenomena is completely determined by its past and in particular by the determination of all positions and momenta at any one moment. In the complete Gibbsian theory, it is still true that with a perfect determination of the multiple time series of the whole universe the knowledge of all positions and momenta at any one moment would determine the entire future. It is only because these are ignored, non-observed coordinates and momenta that the time series with which we actually work take on the sort of mixing property with which we have become familiar in this chapter, in the case of time series derived from the Brownian motion. The great contribution of Heisenberg to physics was the replacement of this still quasi-Newtonian world of Gibbs by one in which the time series can in no way be reduced to an assembly of determinate threads of development in time. In quantum mechanics, the whole past of an individual system does not determine

---

[1] See writings of R. A. Fisher and J. von Neumann.

the future of that system in any absolute way but merely the distribution of possible futures of the system.   The quantities which the classical physics demands for a knowledge of the entire course of a system are not simultaneously observable, except in a loose and approximate way, which nevertheless is sufficiently precise for the needs of the classical physics *over the range of precision where it has been shown experimentally to be applicable.*   The conditions of the observation of a momentum and its corresponding position are incompatible.   To observe the position of a system as precisely as possible, we must observe it with light or electron waves or similar means of high resolving power, or short wavelength.   However, the light has a particle action depending on its frequency only, and to illuminate a body with high-frequency light means to subject it to a change in its momentum which increases with the frequency.   On the other hand, it is low-frequency light that gives the minimum change in the momenta of the particles it illuminates, and this has not a sufficient resolving power to give a sharp indication of positions. Intermediate frequencies of light give a blurred account both of positions and of momenta.   In general, there is no set of observations conceivable which can give us enough information about the past of a system to give us complete information as to its future.

Nevertheless, as in the case of all ensembles of time series, the theory of the amount of information which we have here developed is applicable, and consequently the theory of entropy.   Since, however, we now are dealing with time series with the mixing property, even when our data are as complete as they can be, we find that our system has no absolute potential barriers, and that in the course of time any state of the system can and will transform itself into any other state.   However, the probability of this depends in the long run on the relative probability or measure of the two states.   This turns out to be especially high for states which can be transformed into themselves by a large number of transformations, for states which, in the language of the quantum theorist, have a high internal resonance, or a high quantum degeneracy.   The benzene ring is an example of this, since the two states are equivalent.   This suggests that in a

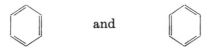

system in which various building blocks may combine themselves intimately in various ways, as when a mixture of amino acids organizes itself into protein chains, a situation where many of these chains are alike and go through a stage of close association with one another may be more stable than one in which they are different. It was suggested by Haldane, in a tentative manner, that this may be the way in which genes and viruses reproduce themselves; and although he has not asserted this suggestion of his with anything like finality, I see no cause not to retain it as a tentative hypothesis. As Haldane himself has pointed out, as no single particle in quantum theory has a perfectly sharp individuality, it is not possible in such a case to say, with more than fragmentary accuracy, which of the two examples of a gene that has reproduced itself in this manner is the master pattern and which is the copy.

This same phenomenon of resonance is known to be very frequently represented in living matter. Szent-Györgyi has suggested its importance in the construction of muscles. Substances with high resonance very generally have an abnormal capacity for storing both energy and information, and such a storage certainly occurs in muscle contraction.

Again, the same phenomena that are concerned in reproduction probably have something to do with the extraordinary specificity of the chemical substances found in a living organism, not only from species to species but even within the individuals of a species. Such considerations may be very important in immunology.

IV

Feedback and Oscillation

A patient comes into a neurological clinic. He is not paralyzed, and he can move his legs when he receives the order. Nevertheless, he suffers under a severe disability. He walks with a peculiar uncertain gait, with eyes downcast on the ground and on his legs. He starts each step with a kick, throwing each leg in succession in front of him. If blindfolded, he cannot stand up, and totters to the ground. What is the matter with him?

Another patient comes in. While he sits at rest in his chair, there seems to be nothing wrong with him. However, offer him a cigarette, and he will swing his hand past it in trying to pick it up. This will be followed by an equally futile swing in the other direction, and this by still a third swing back, until his motion becomes nothing but a futile and violent oscillation. Give him a glass of water, and he will empty it in these swings before he is able to bring it to his mouth. What is the matter with him?

Both of these patients are suffering from one form or another of what is known as *ataxia*. Their muscles are strong and healthy enough, but they are unable to organize their actions. The first patient suffers from *tabes dorsalis*. The part of the spinal cord which ordinarily receives sensations has been damaged or destroyed by the late sequelae of syphilis. The incoming messages are blunted, if they have not totally disappeared. The receptors in the joints and tendons and muscles and the soles of his feet, which ordinarily convey to him the position and state of motion of his legs, send no messages which his central nervous system can pick up and transmit, and for information concerning his posture he is obliged to trust to his

95

eyes and the balancing organs of his inner ear. In the jargon of the
physiologist, he has lost an important part of his proprioceptive or
kinesthetic sense.

The second patient has lost none of his proprioceptive sense. His
injury is elsewhere, in the cerebellum, and he is suffering from what is
known as a cerebellar tremor or purpose tremor. It seems likely
that the cerebellum has some function of proportioning the muscular
response to the proprioceptive input, and if this proportioning is
disturbed, a tremor may be one of the results.

We thus see that for effective action on the outer world it is not
only essential that we possess good effectors, but that the performance
of these effectors be properly monitored back to the central nervous
system, and that the readings of these monitors be properly combined
with the other information coming in from the sense organs to
produce a properly proportioned output to the effectors. Something
quite similar is the case in mechanical systems. Let us consider a
signal tower on a railroad. The signalman controls a number of
levers which turn the semaphore signals on or off and which regulate
the setting of the switches. However, it does not do for him to
assume blindly that the signals and the switches have followed his
orders. It may be that the switches have frozen fast, or that the
weight of a load of snow has bent the signal arms, and that what he
has supposed to be the actual state of the switches and the signals—
his effectors—does not correspond to the orders he has given. To
avoid the dangers inherent in this contingency, every effector, switch
or signal, is attached to a telltale back in the signal tower, which
conveys to the signalman its actual states and performance. This is
the mechanical equivalent of the repeating of orders in the navy,
according to a code by which every subordinate, upon the reception
of an order, must repeat it back to his superior, to show that he has
heard and understood it. It is on such repeated orders that the
signalman must act.

Notice that in this system there is a human link in the chain of
the transmission and return of information: in what we shall from
now on call the chain of feedback. It is true that the signalman is
not altogether a free agent; that his switches and signals are inter-
locked, either mechanically or electrically, and that he is not free to
choose some of the more disastrous combinations. There are,
however, feedback chains in which no human element intervenes.
The ordinary thermostat by which we regulate the heating of a house
is one of these. There is a setting for the desired room temperature;
and if the actual temperature of the house is below this, an apparatus

is actuated which opens the damper, or increases the flow of fuel oil, and brings the temperature of the house up to the desired level.   If, on the other hand, the temperature of the house exceeds the desired level, the dampers are turned off or the flow of fuel oil is slackened or interrupted.   In this way the temperature of the house is kept approximately at a steady level.   Note that the constancy of this level depends on the good design of the thermostat, and that a badly designed thermostat may send the temperature of the house into violent oscillations not unlike the motions of the man suffering from cerebellar tremor.

Another example of a purely mechanical feedback system—the one originally treated by Clerk Maxwell—is that of the governor of a steam engine, which serves to regulate its velocity under varying conditions of load.   In the original form designed by Watt, it consists of two balls attached to pendulum rods and swinging on opposite sides of a rotating shaft.   They are kept down by their own weight or by a spring, and they are swung upward by a centrifugal action dependent on the angular velocity of the shaft.   They thus assume a compromise position likewise dependent on the angular velocity.   This position is transmitted by other rods to a collar about the shaft, which actuates a member which serves to open the intake valves of the cylinder when the engine slows down and the balls fall, and to close them when the engine speeds up and the balls rise.   Notice that the feedback tends to oppose what the system is already doing, and is thus negative.

We have thus examples of negative feedbacks to stabilize temperature and negative feedbacks to stabilize velocity.   There are also negative feedbacks to stabilize position, as in the case of the steering engines of a ship, which are actuated by the angular difference between the position of the wheel and the position of the rudder, and always act so as to bring the position of the rudder into accord with that of the wheel.   The feedback of voluntary activity is of this nature.   We do not will the motions of certain muscles, and indeed we generally do not know which muscles are to be moved to accomplish a given task; we will, say, to pick up a cigarette.   Our motion is regulated by some measure of the amount by which it has not yet been accomplished.

The information fed back to the control center tends to oppose the departure of the controlled from the controlling quantity, but it may depend in widely different ways on this departure.   The simplest control systems are linear: the output of the effector is a linear expression in the input, and when we add inputs, we also add outputs.

The output is read by some apparatus equally linear. This reading is simply subtracted from the input. We wish to give a precise theory of the performance of such a piece of apparatus, and, in particular, of its defective behavior and its breaking into oscillation when it is mishandled or overloaded.

In this book, we have avoided mathematical symbolism and mathematical technique as far as possible, although we have been forced to compromise with them in various places, and in particular in the previous chapter. Here, too, in the rest of the present chapter, we are dealing precisely with those matters for which the symbolism of mathematics is the appropriate language, and we can avoid it only by long periphrases which are scarcely intelligible to the layman, and which are intelligible only to the reader acquainted with mathematical symbolism by virtue of his ability to translate them into this symbolism. The best compromise we can make is to supplement the symbolism by an ample verbal explanation.

Let $f(t)$ be a function of the time t where t runs from $-\infty$ to ∞; that is, let $f(t)$ be a quantity assuming a numerical value for each time t. At any time t, the quantities $f(s)$ are accessible to us when s is less than or equal to t but not when s is greater than t. There are pieces of apparatus, electrical and mechanical, which delay their input by a fixed time, and these yield us, for an input $f(t)$, an output $f(t - \tau)$, where τ is the fixed delay.

We may combine several pieces of apparatus of this kind, yielding us outputs $f(t - \tau_1), f(t - \tau_2), \cdots, f(t - \tau_n)$. We can multiply each of these outputs by fixed quantities, positive or negative. For example, we may use a potentiometer to multiply a voltage by a fixed positive number less than 1, and it is not too difficult to devise automatic balancing devices and amplifiers to multiply a voltage by quantities which are negative or are greater than 1. It is also not difficult to construct simple wiring diagrams of circuits by which we can add voltages continuously, and with the aid of these we may obtain an output

$$\sum_1^n a_k f(t - \tau_k) \tag{4.01}$$

By increasing the number of delays τ_k and suitably adjusting the coefficients a_k, we may approximate as closely as we wish to an output of the form

$$\int_0^\infty a(\tau) f(t - \tau)\, d\tau \tag{4.02}$$

In this expression, it is important to realize that the fact that we are integrating from 0 to ∞, and not from $-\infty$ to ∞, is essential. Otherwise we could use various practical devices to operate on this result and to obtain $f(t + \sigma)$, where σ is positive. This, however, involves the knowledge of the future of $f(t)$; and $f(t)$ may be a quantity, like the coordinates of a streetcar which may turn off one way or the other at a switch, which is not determined by its past. When a physical process *seems* to yield us an operator which converts $f(t)$ to

$$\int_{-\infty}^{\infty} a(\tau)f(t - \tau)\, d\tau \qquad (4.03)$$

where $a(\tau)$ does not effectively vanish for negative values of τ, it means that we have no longer a true operator on $f(t)$, determined uniquely by its past. There are physical cases where this may occur. For example, a dynamical system with no input may go into permanent oscillation, or even oscillation building up to infinity, with an undetermined amplitude. In such a case, the future of the system is not determined by the past, and we may in appearance find a formalism which suggests an operator dependent on the future.

The operation by which we obtain Expression 4.02 from $f(t)$ has two important further properties: (1) it is independent of a shift of the origin of time, and (2) it is linear. The first property is expressed by the statement that if

$$g(t) = \int_{0}^{\infty} \alpha(\tau)f(t - \tau)\, d\tau \qquad (4.04)$$

then

$$g(t + \sigma) = \int_{0}^{\infty} \alpha(\tau)f(t + \sigma - \tau)\, d\tau \qquad (4.05)$$

The second property is expressed by the statement that if

$$g(t) = Af_1(t) + Bf_2(t) \qquad (4.06)$$

then

$$\int_{0}^{\infty} a(\tau)g(t - \tau)\, d\tau$$
$$= A \int_{0}^{\infty} a(\tau)f_1(t - \tau)\, d\tau + B \int_{0}^{\infty} a(\tau)f_2(t - \tau)\, d\tau \quad (4.07)$$

It may be shown that in an appropriate sense *every operator on the past of $f(t)$ which is linear and is invariant under a shift of the origin of time is either of the form of Expression 4.02 or is a limit of a sequence* *of operators of that form.* For example, $f'(t)$ is the result of an operator with these properties when applied to $f(t)$, and

$$f'(t) = \lim_{\epsilon \to 0} \int_0^\infty \frac{1}{\epsilon^2} a\left(\frac{\tau}{\epsilon}\right) f(t - \tau)\, d\tau \qquad (4.08)$$

where

$$a(x) = \begin{cases} 1 & 0 \leqslant x < 1 \\ -1 & 1 \leqslant x < 2 \\ 0 & 2 \leqslant x \end{cases} \qquad (4.09)$$

As we have seen before, the functions e^{zt} are a set of functions $f(t)$ which are particularly important from the point of view of Operator 4.02, since

$$e^{z(t-\tau)} = e^{zt} \cdot e^{-z\tau} \qquad (4.10)$$

and the delay operator becomes merely a multiplier dependent on z. Thus Operator 4.02 becomes

$$e^{zt} \int_0^\infty a(\tau) e^{-z\tau}\, d\tau \qquad (4.11)$$

and is also a multiplication operator dependent on z only. The expression

$$\int_0^\infty a(\tau) e^{-z\tau}\, d\tau = A(z) \qquad (4.12)$$

is said to be *the representation of Operator 4.02 as a function of frequency.* If z is taken as the complex quantity $x + iy$, where x and y are real, this becomes

$$\int_0^\infty a(\tau) e^{-x\tau} e^{-iy\tau}\, d\tau \qquad (4.13)$$

so that by the well-known Schwarz inequality concerning integrals, if $y > 0$ and

$$\int_0^\infty |a(\tau)|^2\, d\tau < \infty \qquad (4.14)$$

we have

$$|A(x + iy)| \leqslant \left[\int_0^\infty |a(\tau)|^2\, d\tau \int_0^\infty e^{-2x\tau}\, d\tau\right]^{\frac{1}{2}}$$

$$= \left[\frac{1}{2x} \int_0^\infty |a(\tau)|^2\, d\tau\right]^{\frac{1}{2}} \qquad (4.15)$$

This means that $A(x + iy)$ is a bounded holomorphic function of a complex variable in every half-plane $x \geqslant \epsilon > 0$, and that the function $A(iy)$ represents in a certain very definite sense the boundary values of such a function.

Let us put

$$u + iv = A(x + iy) \tag{4.16}$$

where u and v are real. The $x + iy$ will be determined as a function (not necessarily single-valued) of $u + iv$. This function will be analytic, though meromorphic, except at the points $u + iv$ corresponding to points $z = x + iy$, where $\partial A(z)/\partial z = 0$. The boundary $x = 0$ will go into the curve with the parametric equation

$$u + iv = A(iy) \qquad (y \text{ real}) \tag{4.17}$$

This new curve may intersect itself any number of times. In general, however, it will divide the plane into two regions. Let us consider the curve (Eq. 4.17) traced in the direction in which y goes from $-\infty$ to ∞. Then if we depart from Eq. 4.17 to the right and follow a continuous course not again cutting Eq. 4.17, we may arrive at certain points. The points which are neither in this set nor on Eq. 4.17 we shall call *exterior points*. The part of the curve (Eq. 4.17) which contains limit points of the exterior points we shall call the *effective boundary*. All other points will be termed *interior points*. Thus in the diagram of Fig. 1, with the boundary drawn in the sense of the arrow, the interior points are shaded and the effective boundary is drawn heavily.

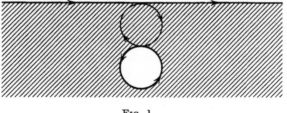

FIG. 1.

The condition that A be bounded in any right half-plane will then tell us that *the point at infinity cannot be an interior point*. It may be a boundary point, although there are certain very definite restrictions on the character of the type of boundary point it may be. These concern the "thickness" of the set of interior points reaching out to infinity.

Now we come to the problem of the mathematical expression of the problem of linear feedback.   Let the control flow chart—*not* the wiring diagram—of such a system be as shown in Fig. 2.   Here

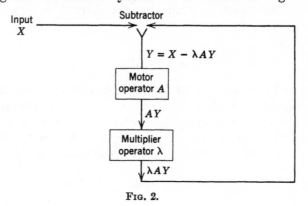

Fɪɢ. 2.

the input of the motor is Y, which is the difference between the original input X and the output of the multiplier, which multiplies the power output AY of the motor by the factor λ.   Thus

$$Y = X - \lambda A Y \tag{4.18}$$

and

$$Y = \frac{X}{1 + \lambda A} \tag{4.19}$$

so that the motor output is

$$A Y = X \frac{A}{1 + \lambda A} \tag{4.20}$$

The operator produced by the whole feedback mechanism is then $A/(1 + \lambda A)$.   *This will be infinite when and only when* $A = -1/\lambda$. *The diagram* (Eq. 4.17) *for this new operator will be*

$$u + iv = \frac{A(iy)}{1 + \lambda A(iy)} \tag{4.21}$$

and ∞ will be an interior point of this when and only when $-1/\lambda$ is an interior point of Eq. 4.17.

In this case, a feedback with a multiplier λ will certainly produce something catastrophic, and as a matter of fact the catastrophe will be that the system will go into unrestrained and increasing oscillation. If, on the other hand, the point $-1/\lambda$ is an exterior point, it may be shown that there will be no difficulty, and the feedback is stable. If $-1/\lambda$ is on the effective boundary, a more elaborate discussion is necessary. Under most circumstances, the system may go into an oscillation of an amplitude which does not increase.

It is perhaps worth considering several operators A and the ranges of feedback which are admissible under them. We shall consider not only the operations of Expression 4.02 but also their limits, assuming that the same argument will apply to these.

If the operator A corresponds to the differential operator, $A(z) = z$, as y goes from $-\infty$ to ∞, $A(y)$ does the same, and the interior points are the points interior to the right half-plane. The point $-1/\lambda$ is always an exterior point, and any amount of feedback is possible. If

$$A(z) = \frac{1}{1 + kz} \tag{4.22}$$

the curve (Eq. 4.17) is

$$u + iv = \frac{1}{1 + kiy} \tag{4.23}$$

or

$$u = \frac{1}{1 + k^2y^2}, \qquad v = \frac{-ky}{1 + k^2y^2} \tag{4.24}$$

which we may write

$$u^2 + v^2 = u \tag{4.25}$$

This is a circle with radius $1/2$, and center at $(1/2, 0)$. It is described in the clockwise sense, and the interior points are those which we should ordinarily consider interior. In this case too, the admissible feedback is unlimited, as $-1/\lambda$ is always outside the circle. The $a(t)$ corresponding to this operator is

$$a(t) = e^{-t/k}/k \tag{4.26}$$

Again, let

$$A(z) = \left(\frac{1}{1 + kz}\right)^2 \tag{4.27}$$

Then Eq. 4.17 is

$$u + iv = \left(\frac{1}{1 + kiy}\right)^2 = \frac{(1 - kiy)^2}{(1 + k^2y^2)^2} \tag{4.28}$$

and

$$u = \frac{1 - k^2y^2}{(1 + k^2y^2)^2}, \qquad v = \frac{-2ky}{(1 + k^2y^2)^2} \tag{4.29}$$

This yields

$$u^2 + v^2 = \frac{1}{(1 + k^2y^2)^2} \tag{4.30}$$

or

$$y = \frac{-v}{(u^2 + v^2)2k} \tag{4.31}$$

Then

$$u = (u^2 + v^2)\left[1 - \frac{k^2v^2}{4k^2(u^2 + v^2)^2}\right] = (u^2 + v^2) - \frac{v^2}{4(u^2 + v^2)} \tag{4.32}$$

In polar coordinates, if $u = \rho \cos \phi$, $v = \rho \sin \phi$, this becomes

$$\rho \cos \phi = \rho^2 - \frac{\sin^2\phi}{4} = \rho^2 - \frac{1}{4} + \frac{\cos^2\phi}{4} \tag{4.33}$$

or

$$\rho - \frac{\cos \phi}{2} = \pm \frac{1}{2} \tag{4.34}$$

That is,

$$\rho^{1/2} = -\sin \frac{\phi}{2}, \qquad \rho^{1/2} = \cos \frac{\phi}{2} \tag{4.35}$$

It can be shown that these two equations represent only one curve, a cardioid with vertex at the origin and cusp pointing to the right. The interior of this curve will contain no point of the negative real axis, and, as in the previous case, the admissible amplification is unlimited. Here the operator $a(t)$ is

$$a(t) = \frac{t}{k^2} e^{-t/k} \tag{4.36}$$

Let

$$A(z) = \left(\frac{1}{1 + kz}\right)^3 \tag{4.37}$$

Let ρ and ϕ be defined as in the last case. Then

$$\rho^{1/3} \cos \frac{\phi}{3} + i\rho^{1/3} \sin \frac{\phi}{3} = \frac{1}{1 + kiy} \tag{4.38}$$

As in the first case, this will give us

$$\rho^{2/3} \cos^2 \frac{\phi}{3} + \rho^{2/3} \sin^2 \frac{\phi}{3} = \rho^{1/3} \cos \frac{\phi}{3} \tag{4.39}$$

That is,

$$\rho^{1/3} = \cos \frac{\phi}{3} \tag{4.40}$$

which is a curve of the shape of Fig. 3.   The shaded region represents

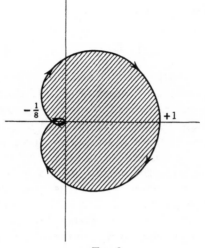

Fig. 3.

the interior points.   All feedback with coefficient exceeding 1/8 is impossible.   The corresponding $a(t)$ is

$$a(t) = \frac{t^2}{2k^3} e^{-t/k} \tag{4.41}$$

Finally, let our operator corresponding to A be a simple delay of T units of time.   Then

$$A(z) = e^{-Tz} \tag{4.42}$$

Then

$$u + iv = e^{-Tiy} = \cos Ty - i \sin Ty \tag{4.43}$$

The curve (Eq. 4.17) will be the unit circle about the origin, described in a clockwise sense about the origin with unit velocity. The inside of this curve will be the inside in the ordinary sense, and the limit of feedback intensity will be 1.

There is one very interesting conclusion to be drawn from this. It is possible to compensate for the operator $1/(1 + kz)$ by an arbitrarily heavy feedback, which will give us an $A/(1 + \lambda A)$ as near to 1 as we wish for as large a frequency range as we wish.   It is thus possible to compensate for three successive operators of this sort by three—or even two—successive feedbacks.   It is not, however, possible to compensate as closely as we wish for an operator

$1/(1 + kz)^3$, which is the resultant of the composition of three operators $1/(1 + kz)$ in cascade, by a single feedback. The operator $1/(1 + kz)^3$ may also be written

$$\frac{1}{2k^2} \frac{d^2}{dz^2} \frac{1}{1 + kz} \qquad (4.44)$$

and may be regarded as the limit of the additive composition of three operators with first-degree denominators. It thus appears that a sum of different operators, each of which may be compensated as well as we wish by a single feedback, cannot itself be so compensated.

In the important book of MacColl, we have an example of a complicated system which can be stabilized by two feedbacks but not by one. It concerns the steering of a ship by a gyrocompass. The angle between the course set by the quartermaster and that shown by the compass expresses itself in the turning of the rudder, which, in view of the headway of the ship, produces a turning moment which serves to change the course of the ship in such a way as to decrease the difference between the set course and the actual course. If this is done by a direct opening of the valves of one steering engine and closing of the valves of the other in such a way that the turning velocity of the rudder is proportional to the deviation of the ship from this course, let us note that the angular position of the rudder is roughly proportional to the turning moment of the ship and thus to its angular acceleration. Hence the amount of turning of the ship is proportional with a negative factor to the third derivative of the deviation from the course, and the operation which we have to stabilize by the feedback from the gyrocompass is kz^3, where k is positive. We thus get for our curve (Eq. 4.17)

$$u + iv = -kiy^3 \qquad (4.45)$$

and, as the left half-plane is the interior region, no servomechanism whatever will stabilize the system.

In this account, we have slightly oversimplified the steering problem. Actually there is a certain amount of friction, and the force turning the ship does not determine the acceleration. Instead, if θ is the angular position of the ship and ϕ that of the rudder with respect to the ship, we have

$$\frac{d^2\theta}{dt^2} = c_1\phi - c_2\frac{d\theta}{dt} \qquad (4.46)$$

and

$$u + iv = -k_1iy^3 - k_2y^2 \qquad (4.47)$$

This curve may be written

$$v^2 = -k_3 u^3 \qquad (4.48)$$

which still cannot be stabilized by any feedback. As y goes from $-\infty$ to ∞, u goes from ∞ to $-\infty$, and the *inside* of the curve is to the left.

If, on the other hand, the *position* of the rudder is proportional to the deviation of the course, the operator to be stabilized by feedback is $k_1 z^2 + k_2 z$, and Eq. 4.17 becomes

$$u + iv = -k_1 y^2 + k_2 iy \qquad (4.49)$$

This curve may be written

$$v^2 = -k_3 u \qquad (4.50)$$

but in this case, as y goes from $-\infty$ to ∞, so does v, and the curve is described from $y = -\infty$ to $y = \infty$. In this case, the *outside* of the curve is to the left, and unlimited amount of amplification is possible.

To achieve this we may employ another stage of feedback. If we regulate the position of the valves of the steering engine, not by the discrepancy between the actual and the desired course but by the *difference* between this quantity and the angular position of the rudder, we shall keep the angular position of the rudder as nearly proportional to the ship's deviation from true course as we wish, if we allow a large enough feedback—that is, if we open the valves wide enough. This double feedback system of control is in fact the one usually adopted for the automatic steering of ships by means of the gyrocompass.

In the human body, the motion of a hand or a finger involves a system with a large number of joints. The output is an additive vectorial combination of the outputs of all these joints. We have seen that, in general, a complex additive system like this cannot be stabilized by a single feedback. Correspondingly, the voluntary feedback by which we regulate the performance of a task through the observation of the amount by which it is not yet accomplished needs the backing up of other feedbacks. These we call postural feedbacks, and they are associated with the general maintenance of tone of the muscular system. It is the voluntary feedback which shows a tendency to break down or become deranged in cases of cerebellar injury, for the ensuing tremor does not appear unless the patient tries to perform a voluntary task. This purpose tremor, in which the patient cannot pick up a glass of water without upsetting it, is very different in nature from the tremor of Parkinsonism, or paralysis *agitans*, which appears in its most typical form when the patient is at rest, and indeed often seems to be greatly mitigated when he attempts to perform a specific task.   There are surgeons with Parkinsonism who manage to operate quite efficiently. Parkinsonism is known not to have its origin in a diseased condition of the cerebellum, but to be associated with a pathological focus somewhere in the brain stem.   It is only one of the diseases of the postural feedbacks, and many of these must have their origin in defects of parts of the nervous system situated very differently.   One of the great tasks of physiological cybernetics is to disentangle and isolate loci of the different parts of this complex of voluntary and postural feedbacks.   Examples of component reflexes of this sort are the scratch and the walking reflex.

When feedback is possible and stable, its advantage, as we have already said, is to make performance less dependent on the load. Let us consider that the load changes the characteristic A by dA. The fractional change will be dA/A.   If the operator after feedback is

$$B = \frac{A}{C + A} \tag{4.51}$$

we shall have

$$\frac{dB}{B} = \frac{-d\left(1 + \dfrac{C}{A}\right)}{1 + \dfrac{C}{A}} = \frac{\dfrac{C}{A^2}\, dA}{1 + \dfrac{C}{A}} = \frac{dA}{A}\,\frac{C}{A + C} \tag{4.52}$$

Thus feedback serves to diminish the dependence of the system on the characteristic of the motor, and serves to stabilize it, for all frequencies for which

$$\left|\frac{A + C}{C}\right| > 1 \tag{4.53}$$

This is to say that the entire boundary between interior and exterior points must lie inside the circle of radius C about the point $-C$. This will not even be true in the first of the cases we have discussed. The effect of a heavy negative feedback, if it is at all stable, will be to increase the stability of the system for low frequencies, but generally at the expense of its stability for some high frequencies. There are many cases in which even this degree of stabilization is advantageous.

A very important question which arises in connection with oscillations due to an excessive amount of feedback is that of the frequency of incipient oscillation.   This is determined by the value of y in the

iy corresponding to the point of the boundary of the inside and outside regions of Eq. 4.17 lying furthest to the left on the negative u-axis. The quantity y is of course of the nature of a frequency.

We have now come to the end of an elementary discussion of linear oscillations, studied from the point of view of feedback. A linear oscillating system has certain very special properties which characterize its oscillations. One is that when it oscillates, it always *can* and very generally—in the absence of independent simultaneous oscillations—*does* oscillate in the form

$$A \sin (Bt + C)e^{Dt} \qquad (4.54)$$

The existence of a periodic non-sinusoidal oscillation is always a suggestion at least that the variable observed is one in which the system is not linear. In some cases, but in very few, the system may be rendered linear again by a new choice of the independent variable.

Another very significant difference between linear and non-linear oscillations is that in the first the amplitude of oscillation is completely independent of the frequency; while in the latter, there is generally only one amplitude, or at most a discrete set of amplitudes, for which the system will oscillate at a given frequency, as well as a discrete set of frequencies for which the system will oscillate. This is well illustrated by the study of what happens in an organ pipe. There are two theories of the organ pipe—a cruder linear theory, and a more precise non-linear theory. In the first, the organ pipe is treated as a conservative system. No question is asked about how the pipe came to oscillate, and the level of oscillation is completely indeterminate. In the second theory, the oscillation of the organ pipe is considered as dissipating energy, and this energy is considered to have its origin in the stream of air across the lip of the pipe. There is indeed a theoretical steady-state flow of air across the lip of the pipe which does not interchange any energy with any of the modes of oscillation of the pipe, but for certain velocities of air flow this steady-state condition is unstable. The slightest chance deviation from it will introduce an energy input into one or more of the natural modes of linear oscillation of the pipe; and up to a certain point, this motion will actually increase the coupling of the proper modes of oscillation of the pipe with the energy input. The rate of energy input and the rate of energy output by thermal dissipation and otherwise have different laws of growth, but, to arrive at a steady state of oscillation, these two quantities must be identical. Thus the level of the non-linear oscillation is determined just as definitely as its frequency.

The case we have examined is an example of what is known as a relaxation oscillation: a case, that is, where a system of equations invariant under a translation in time leads to a solution periodic—or corresponding to some generalized notion of periodicity—in time, and determinate in amplitude and frequency but not in phase. In the case we have discussed, the frequency of oscillation of the system is close to that of some loosely coupled, nearly linear part of the system. B. van der Pol, one of the chief authorities on relaxation oscillations, has pointed out that this is not always the case, and that there are in fact relaxation oscillations where the predominating frequency is not near the frequency of linear oscillation of any part of the system. An example is given by a stream of gas flowing into a chamber open to the air and in which a pilot light is burning: when the concentration of gas in the air reaches a certain critical value, the system is ready to explode under ignition by the pilot light, and the time it takes for this to happen depends only on the rate of flow of the coal gas, the rate at which air seeps in and the products of combustion seep out, and the percentage composition of an explosive mixture of coal gas and air.

In general, non-linear systems of equations are hard to solve. There is, however, a specially tractable case, in which the system differs only slightly from a linear system, and the terms which distinguish it change so slowly that they may be considered substantially constant over a period of oscillation. In this case, we may study the non-linear system as if it were a linear system with slowly varying parameters. Systems which may be studied this way are said to be perturbed secularly, and the theory of secularly perturbed systems plays a most important role in gravitational astronomy.

It is quite possible that some of the physiological tremors may be treated somewhat roughly as secularly perturbed linear systems. We can see quite clearly in such a system why the steady-state amplitude level may be just as determinate as the frequency. Let one element in such a system be an amplifier whose gain decreases as some long-time average of the input of such a system increases. Then as the oscillation of the system builds up, the gain may be reduced until a state of equilibrium is reached.

Non-linear systems of relaxation oscillations have been studied in some cases by methods developed by Hill and Poincaré.[1] The classical cases for the study of such oscillations are those in which

[1] Poincaré, H., *Les Méthodes Nouvelles de la Mécanique Céleste*, Gauthier-Villars et fils, Paris, 1892–1899.

the equations of the systems are of a different nature; especially where these differential equations are of low order. There is not, as far as I know, any comparable adequate study of the corresponding integral equations when the system depends for its future behavior on its entire past behavior. However, it is not hard to sketch out the form such a theory should take, especially when we are looking only for periodic solutions. In this case, the slight modification of the constants of the equation should lead to a slight, and therefore nearly linear, modification of the equations of motion. For example, let $Op[f(t)]$ be a function of t which results from a non-linear operation on $f(t)$, and which is affected by a translation. Then the variation of $Op[f(t)]$, $\delta Op[f(t)]$ corresponding to a variational change $\delta f(t)$ in $f(t)$ and a known change in the dynamics of the system, is linear but not homogeneous in $\delta f(t)$, though not linear in $f(t)$. If we now know a solution $f(t)$ of

$$Op[f(t)] = 0 \qquad (4.55)$$

and we change the dynamics of the system, we obtain a linear non-homogeneous equation for $\delta f(t)$. If

$$f(t) = \sum_{-\infty}^{\infty} a_n e^{in\lambda t} \qquad (4.56)$$

and $f(t) + \delta f(t)$ is also periodic, being of the form

$$f(t) + \delta f(t) = \sum_{-\infty}^{\infty} (a_n + \delta a_n) e^{in(\lambda + \delta \lambda)t} \qquad (4.57)$$

then

$$\delta f(t) = \sum_{-\infty}^{\infty} \delta a_n e^{i\lambda nt} + \sum_{-\infty}^{\infty} a_n e^{i\lambda nt} in\delta\lambda t \qquad (4.58)$$

The linear equations for $\delta f(t)$ will have all coefficients developable into series in $e^{i\lambda nt}$, since $f(t)$ can itself be developed in this form. We shall thus obtain an infinite system of linear non-homogeneous equations in $\delta a_n + a_n$, $\delta\lambda$, and λ, and this system of equations may be solvable by the methods of Hill. In this case, it is at least conceivable that by starting with a linear equation (non-homogeneous) and gradually shifting the constraints we may arrive at a solution of a very general type of non-linear problem in relaxation oscillations. This work, however, lies in the future.

To a certain extent, the feedback systems of control discussed in this chapter and the compensation systems discussed in the previous one are competitors. They both serve to bring the complicated input-output relations of an effector into a form approaching a simple proportionality. The feedback system, as we have seen, does more than this, and has a performance relatively independent of the characteristic and changes of characteristic of the effector used. The relative usefulness of the two methods of control thus depends on the constancy of the characteristic of the effector. It is natural to suppose that cases arise in which it is advantageous to combine the two methods. There are various ways of doing this. One of the most simple is that illustrated in the diagram of Fig. 4.

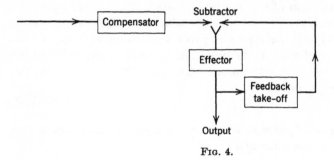

FIG. 4.

In this, the entire feedback system may be regarded as a larger effector, and no new point arises, except that the compensator must be arranged to compensate what is in some sense the average characteristic of the feedback system. Another type of arrangement is shown in Fig. 5.

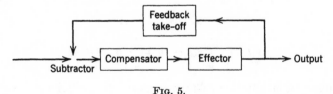

FIG. 5.

Here the compensator and effector are combined into one larger effector. This change will in general alter the maximum feedback admissible, and it is not easy to see how it can ordinarily be made to increase that level to an important extent. On the other hand, for the same feedback level, it will most definitely improve the performance of the system. If, for example, the effector has an essentially lagging characteristic, the compensator will be an anticipator or predictor, designed for its statistical ensemble of inputs. Our feedback, which we may call an anticipatory feedback, will tend to hurry up the action of the effector mechanism.

Feedbacks of this general type are certainly found in human and animal reflexes. When we go duck shooting, the error which we try to minimize is not that between the position of the gun and the actual position of the target but that between the position of the gun and the anticipated position of the target. Any system of anti-aircraft fire control must meet the same problem. The conditions of stability and effectiveness of anticipatory feedbacks need a more thorough discussion than they have yet received.

Another interesting variant of feedback systems is found in the way in which we steer a car on an icy road. Our entire conduct of driving depends on a knowledge of the slipperiness of the road surface, that is, on a knowledge of the performance characteristics of the system car–road. If we wait to find this out by the ordinary performance of the system, we shall discover ourselves in a skid before we know it. We thus give to the steering wheel a succession of small, fast impulses, not enough to throw the car into a major skid but quite enough to report to our kinesthetic sense whether the car is in danger of skidding, and we regulate our method of steering accordingly.

This method of control, which we may call *control by informative feedback*, is not difficult to schematize into a mechanical form and may well be worth while employing in practice. We have a compensator for our effector, and this compensator has a characteristic which may be varied from outside. We superimpose on the incoming message a weak high-frequency input and take off the output of the effector a partial output of the same high frequency, separated from the rest of the output by an appropriate filter. We explore the amplitude-phase relations of the high-frequency output to the input in order to obtain the performance characteristics of the effector. On the basis of this, we modify in the appropriate sense the characteristics of the compensator. The flow chart of the system is much as in the diagram of Fig. 6.

The advantages of this type of feedback are that the compensator may be adjusted to give stability for every type of constant load; and that, if the characteristic of the load changes slowly enough, in what we have called a secular manner, in comparison with the changes of the original input, and if the reading of the load condition is accurate, the system has no tendency to go into oscillation. There are very many cases where the change of load is secular in this manner. For example, the frictional load of a gun turret depends on the stiffness of the grease, and this again on the temperature; but this stiffness will not change appreciably in a few swings of the turret.

Of course, this informative feedback will work well only if the characteristics of the load at high frequencies are the same as, or give a good indication of, its characteristics at low frequencies.

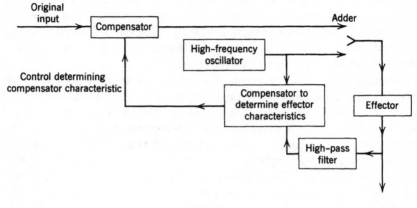

FIG. 6.

This will often be the case if the character of the load, and hence of the effector, contains a relatively small number of variable parameters.

This informative feedback and the examples we have given of feedback with compensators are only particular cases of what is a very complicated theory, and a theory as yet imperfectly studied. The whole field is undergoing a very rapid development. It deserves much more attention in the near future.

Before we end this chapter, we must not forget another important physiological application of the principle of feedback. A great group of cases in which some sort of feedback is not only exemplified in physiological phenomena but is absolutely essential for the continuation of life is found in what is known as *homeostasis*. The conditions under which life, especially healthy life, can continue in the higher animals are quite narrow. A variation of one-half degree centigrade in the body temperature is generally a sign of illness, and a permanent variation of five degrees is scarcely consistent with life. The osmotic pressure of the blood and its hydrogen-ion concentration must be held within strict limits. The waste products of the body must be excreted before they rise to toxic concentrations. Beside all these, our leucocytes and our chemical defenses against infection must be kept at adequate levels; our heart rate and blood pressure must neither be too high nor too low; our sex cycle must conform to the racial needs of reproduction; our calcium metabolism must be such as neither to soften our bones nor to calcify our tissues; and so on. In short, our inner economy must contain an assembly of thermostats, automatic hydrogen-ion-concentration controls, governors, and the like, which would be adequate for a great chemical plant. These are what we know collectively as our homeostatic mechanism.

Our homeostatic feedbacks have one general difference from our voluntary and our postural feedbacks: they tend to be slower. There are very few changes in physiological homeostasis—not even cerebral anemia—that produce serious or permanent damage in a small fraction of a second. Accordingly, the nerve fibers reserved for the processes of homeostasis—the sympathetic and para-sympathetic systems—are often non-myelinated and are known to have a considerably slower rate of transmission than the myelinated fibers. The typical effectors of homeostasis—smooth muscles and glands—are likewise slow in their action compared with striped muscles, the typical effectors of voluntary and postural activity. Many of the messages of the homeostatic system are carried by non-nervous channels—the direct anastomosis of the muscular fibers of the heart, or chemical messengers such as the hormones, the carbon dioxide content of the blood, etc.; and, except in the case of the heart muscle, these too are generally slower modes of transmission than myelinated nerve fibers.

Any complete textbook on cybernetics should contain a thorough detailed discussion of homeostatic processes, many individual cases of which have been discussed in the literature with some detail.[1] However, this book is rather an introduction to the subject than a compendious treatise, and the theory of homeostatic processes involves rather too detailed a knowledge of general physiology to be in place here.

[1] Cannon, W., *The Wisdom of the Body*, W. W. Norton & Company, Inc., New York, 1932; Henderson, L. J., *The Fitness of the Environment*, The Macmillan Company, New York, 1913.

V

Computing Machines
and the Nervous System

Computing machines are essentially machines for recording numbers, operating with numbers, and giving the result in numerical form. A very considerable part of their cost, both in money and in the effort of construction, goes to the simple problem of recording numbers clearly and accurately. The simplest mode of doing this seems to be on a uniform scale, with a pointer of some sort moving over this. If we wish to record a number with an accuracy of one part in n, we have to assure that in each region of the scale the pointer assumes the desired position within this accuracy. That is, for an amount of information $\log_2 n$, we must finish each part of the movement of the pointer with this degree of accuracy, and the cost will be of the form An, where A is not too far from a constant. More precisely, since if $n - 1$ regions are accurately established, the remaining region will also be determined accurately, the cost of recording an amount of information I will be about

$$(2^I - 1)A \qquad (5.01)$$

Now let us divide this information over two scales, each marked less accurately. The cost of recording this information will be about

$$2(2^{I/2} - 1)A \qquad (5.02)$$

If the information be divided among N scales, the approximate cost will be

$$N(2^{I/N} - 1)A \qquad (5.03)$$

This will be a minimum when

$$2^{I/N} - 1 = \frac{I}{N} 2^{I/N} \log 2 \qquad (5.04)$$

or if we put

$$\frac{I}{N} \log 2 = x \qquad (5.05)$$

when

$$x = \frac{e^x - 1}{e^x} = 1 - e^{-x} \qquad (5.06)$$

This will occur when and only when $x = 0$, or $N = \infty$. That is, N should be as large as possible to give the lowest cost for the storage of information. Let us remember that $2^{I/N}$ must be an integer, and that 1 is not a significant value, as we then have an infinite number of scales each containing no information. The best significant value for $2^{I/N}$ is 2, in which case we record our number on a number of independent scales, each divided into two equal parts. In other words, we represent our numbers in the binary system on a number of scales in which all that we know is that a certain quantity lies in one or the other of two equal portions of the scale, and in which the probability of an imperfect knowledge as to which half of the scale contains the observation is made vanishingly small. In other words, we represent a number ν in the form

$$\nu = \nu_0 + \frac{1}{2} \nu_1 + \frac{1}{2^2} \nu_2 + \cdots + \frac{1}{2_n} \nu_n + \cdots \qquad (5.07)$$

where every ν_n is either 1 or 0.

There exist at present two great types of computing machines: those like the Bush differential analyzer,[1] which are known as *analogy machines*, where the data are represented by measurements on some continuous scale, so that the accuracy of the machine is determined by the accuracy of construction of the scale; and those, like the ordinary desk adding and multiplying machine, which we call *numerical machines*, where the data are represented by a set of choices among a number of contingencies, and the accuracy is determined by the sharpness with which the contingencies are distinguished, the number of alternative contingencies presented at every choice, and the number of choices given. We see that for highly accurate work, at any rate, the numerical machines are preferable, and above all, those numerical machines constructed on the binary scale, in which

---

[1] *Journal of the Franklin Institute*, various papers, 1930 on.

the number of alternatives presented at each choice is two.   Our use of machines on the decimal scale is conditioned merely by the historical accident that the scale of ten, based on our fingers and thumbs, was already in use when the Hindus made the great discovery of the importance of the zero and the advantage of a positional system of notation.   It is worth retaining when a large part of the work done with the aid of the machine consists in transcribing onto the machine numbers in the conventional decimal form, and in taking off the machine numbers which must be written in the same conventional form.

This is, in fact, the use of the ordinary desk computing machine, as employed in banks, in business offices, and in many statistical laboratories.   It is not the way that the larger and more automatic machines are best to be employed; in general, any computing machine is used because machine methods are faster than hand methods.   In any combined use of means of computation, as in any combination of chemical reactions, it is the slowest which gives the order of magnitude of the time constants of the entire system.   It is thus advantageous, as far as possible, to remove the human element from any elaborate chain of computation and to introduce it only where it is absolutely unavoidable, at the very beginning and the very end.   Under these conditions, it pays to have an instrument for the change of the scale of notation, to be used initially and finally in the chain of computations, and to perform all intermediate processes on the binary scale.

The ideal computing machine must then have all its data inserted at the beginning, and must be as free as possible from human interference to the very end.   This means that not only must the numerical data be inserted at the beginning, but also all the rules for combining them, in the form of instructions covering every situation which may arise in the course of the computation.   Thus the computing machine must be a logical machine as well as an arithmetic machine and must combine contingencies in accordance with a systematic algorithm.   While there are many algorithms which *might* be used for combining contingencies, the simplest of these is known as the algebra of logic *par excellence*, or the Boolean algebra. This algorithm, like the binary arithmetic, is based on the dichotomy, the choice between *yes* and *no*, the choice between being in a class and outside.   The reasons for its superiority to other systems are of the same nature as the reasons for the superiority of the binary arithmetic over other arithmetics.

Thus all the data, numerical or logical, put into the machine are in the form of a set of choices between two alternatives, and all the operations on the data take the form of making a set of new choices depend on a set of old choices.   When I add two one-digit numbers, A and B, I obtain a two-digit number commencing with 1, if A and B are both 1, and otherwise with 0.   The second digit is 1 if $A \neq B$, and is otherwise 0.   The addition of numbers of more than one digit follows similar but more complicated rules.   Multiplication in the binary system, as in the decimal, may be reduced to the multiplication table and the addition of numbers, and the rules for multiplication for binary numbers take on the peculiarly simple form given by the table

$$
\begin{array}{c|cc}
\times & 0 & 1 \\
\hline
0 & 0 & 0 \\
1 & 0 & 1
\end{array}
\tag{5.08}
$$

Thus multiplication is simply a method to determine a set of new digits when old digits are given.

On the logical side, if O is a negative and I a positive decision, every operator can be derived from three: *negation*, which transforms I into O and O into I; *logical addition*, with the table

$$
\begin{array}{c|cc}
\oplus & O & I \\
\hline
O & O & I \\
I & I & I
\end{array}
\tag{5.09}
$$

and *logical multiplication*, with the same table as the numerical multiplication of the (1, 0) system, namely,

$$
\begin{array}{c|cc}
\odot & O & I \\
\hline
O & O & O \\
I & O & I
\end{array}
\tag{5.10}
$$

That is, every contingency which may arise in the operation of the machine simply demands a new set of choices of contingencies I and O, depending according to a fixed set of rules on the decisions already made.   In other words, the structure of the machine is that of a bank of relays, capable each of two conditions, say "on" and "off"; while at each stage the relays assume each a position dictated by the positions of some or all the relays of the bank at a previous stage of operation.   These stages of operation may be definitely "clocked" from some central clock or clocks, or the action of each relay may be held up until all the relays which should have acted earlier in the process have gone through all the steps called for.

The relays used in a computing machine may be of very varied character. They may be purely mechanical, or they may be electro-mechanical, as in the case of a solenoidal relay, in which the armature will remain in one of two possible positions of equilibrium until an appropriate impulse pulls it to the other side. They may be purely electrical systems with two alternative positions of equilibrium, either in the form of gas-filled tubes, or, what is much more rapid, in the form of high-vacuum tubes. The two possible states of a relay system may both be stable in the absence of outside interference, or only one may be stable, while the other is transitory. Always in the second case and generally in the first case, it will be desirable to have special apparatus to retain an impulse which is to act at some future time, and to avoid the clogging up of the system which will ensue if one of the relays does nothing but repeat itself indefinitely. However, we shall have more to say concerning this question of memory later.

It is a noteworthy fact that the human and animal nervous systems, which are known to be capable of the work of a computation system, contain elements which are ideally suited to act as relays. These elements are the so-called *neurons* or nerve cells. While they show rather complicated properties under the influence of electrical currents, in their ordinary physiological action they conform very nearly to the "all-or-none" principle; that is, they are either at rest, or when they "fire" they go through a series of changes almost independent of the nature and intensity of the stimulus. There is first an active phase, transmitted from one end to the other of the neuron with a definite velocity, to which there succeeds a refractory period during which the neuron is either incapable of being stimulated, or at any rate is not capable of being stimulated by any normal, physiological process. At the end of this effective refractory period, the nerve remains inactive, but may be stimulated again into activity.

Thus the nerve may be taken to be a relay with essentially two states of activity: firing and repose. Leaving aside those neurons which accept their messages from free endings or sensory end organs, each neuron has its message fed into it by other neurons at points of contact known as *synapses*. For a given outgoing neuron, these vary in number from a very few to many hundred. It is the state of the incoming impulses at the various synapses, combined with the antecedent state of the outgoing neuron itself, which determines whether it will fire or not. If it is neither firing nor refractory, and the number of incoming synapses which "fire" within a certain very short fusion interval of time exceeds a certain threshold, then the neuron will fire after a known, fairly constant synaptic delay.

This is perhaps an oversimplification of the picture: the "threshold" may not depend simply on the number of synapses but on their "weight" and their geometrical relations to one another with respect to the neuron into which they feed; and there is very convincing evidence that there exist synapses of a different nature, the so-called "inhibitory synapses," which either completely prevent the firing of the outgoing neuron or at any rate raise its threshold with respect to stimulation at the ordinary synapses.   What is pretty clear, however, is that some definite combinations of impulses on the incoming neurons having synaptic connections with a given neuron will cause it to fire, while others will not cause it to fire.   This is not to say that there may not be other, non-neuronic influences, perhaps of a humoral nature, which produce slow, secular changes tending to vary that pattern of incoming impulses which is adequate for firing.

A very important function of the nervous system, and, as we have said, a function equally in demand for computing machines, is that of *memory*, the ability to preserve the results of past operations for use in the future.   It will be seen that the uses of the memory are highly various, and it is improbable that any single mechanism can satisfy the demands of all of them.   There is first the memory which is necessary for the carrying out of a current process, such as a multiplication, in which the intermediate results are of no value when once the process is completed, and in which the operating apparatus should then be released for further use.   Such a memory should record quickly, be read quickly, and be erased quickly.   On the other hand, there is the memory which is intended to be part of the files, the permanent record, of the machine or the brain, and to contribute to the basis of all its future behavior, at least during a single run of the machine.   Let it be remarked parenthetically that an important difference between the way in which we use the brain and the machine is that the machine is intended for many successive runs, either with no reference to each other, or with a minimal, limited reference, and that it can be cleared between such runs; while the brain, in the course of nature, never even approximately clears out its past records.   Thus the brain, under normal circumstances, is not the complete analogue of the computing machine but rather the analogue of a single run on such a machine.   We shall see later that this remark has a deep significance in psychopathology and in psychiatry.

To return to the problem of memory, a very satisfactory method for constructing a short-time memory is to keep a sequence of impulses traveling around a closed circuit until this circuit is cleared by intervention from outside. There is much reason to believe that this happens in our brains during the retention of impulses, which occurs over what is known as the specious present. This method has been imitated in several devices which have been used in computing machines, or at least suggested for such a use. There are two conditions which are desirable in such a retentive apparatus: the impulse should be transmitted in a medium in which it is not too difficult to achieve a considerable time lag; and before the errors inherent in the instrument have blurred it too much, the impulse should be reconstructed in a form as sharp as possible. The first condition tends to rule out delays produced by the transmission of light, or even, in many cases, by electric circuits, while it favors the use of one form or another of elastic vibrations; and such vibrations have actually been employed for this purpose in computing machines. If electric circuits are used for delay purposes, the delay produced at every stage is relatively short; or, as in all pieces of linear apparatus, the deformation of the message is cumulative and very soon becomes intolerable. To avoid this, a second consideration comes into play; we must insert somewhere in the cycle a relay which does not serve to repeat the form of the incoming message but rather to trigger off a new message of prescribed form. This is done very easily in the nervous system, where indeed all transmission is more or less of a trigger phenomenon. In the electrical industry, pieces of apparatus for this purpose have long been known and have been used in connection with telegraph circuits. They are known as *telegraph-type repeaters*. The great difficulty of using them for memories of long duration is that they have to function without a flaw over an enormous number of consecutive cycles of operation. Their success is all the more remarkable: in a piece of apparatus designed by Mr. Williams of the University of Manchester, a device of this sort with a unit delay of the order of a hundredth of a second has continued in successful operation for several hours. What makes this more remarkable is that this apparatus was not used merely to preserve a single decision, a single "yes" or "no," but a matter of thousands of decisions.

Like other forms of apparatus intended to retain a large number of decisions, this works on the scanning principle. One of the simplest modes of storing information for a relatively short time is as the charge on a condenser; and when this is supplemented by a telegraph-type repeater, it becomes an adequate method of storage. To use to the best advantage the circuit facilities attached to such a storage system, it is desirable to be able to switch successively and very rapidly from one condenser to another. The ordinary means of doing this involve mechanical inertia, and this is never consistent with very high speeds. A much better way is the use of a large number of condensers, in which one plate is either a small piece of metal sputtered in to a dielectric, or the imperfectly insulating surface of the dielectric itself, while one of the connectors to these condensers is a pencil of cathode rays moved by the condensers and magnets of a sweep circuit over a course like that of a plough in a ploughed field. There are various elaborations of this method, which indeed was employed in a somewhat different way by the Radio Corporation of America before it was used by Mr. Williams.

These last-named methods for storing information can hold a message for quite an appreciable time, if not for a period comparable with a human lifetime. For more permanent records, there is a wide variety of alternatives among which we can choose. Leaving out such bulky, slow, and unerasable methods as the use of punched cards and punched tape, we have magnetic tape, together with its modern refinements, which have largely eliminated the tendency of messages on this material to spread; phosphorescent substances; and above all, photography. Photography is indeed ideal for the permanence and detail of its records, ideal again from the point of view of the shortness of exposure needed to record an observation. It suffers from two grave disadvantages: the time needed for development, which has been reduced to a few seconds, but is still not small enough to make photography available for a short-time memory; and (at present [1947]) the fact that a photographic record is not subject to rapid erasure and the rapid implanting of a new record. The Eastman people have been working on just these problems, which do not seem to be necessarily insoluble, and it is possible that by this time they have found the answer.

Very many of the methods of storage of information already considered have an important physical element in common. They seem to depend on systems with a high degree of quantum degeneracy, or, in other words, with a large number of modes of vibration of the same frequency. This is certainly true in the case of ferromagnetism, and is also true in the case of materials with an exceptionally high dielectric constant, which are thus especially valuable for use in condensers for the storage of information. Phosphorescence as well is a phenomenon associated with a high quantum degeneracy, and the same sort of effect makes its appearance in the photographic process, where many of the substances which act as developers seem to have a great deal of internal resonance. Quantum degeneracy appears to be associated with the ability to make small causes produce appreciable and stable effects. We have already seen in Chapter II that substances with high quantum degeneracy appear to be associated with many of the problems of metabolism and reproduction. It is probably not an accident that here, in a non-living environment, we find them associated with a third fundamental property of living matter: the ability to receive and organize impulses and to make them effective in the outer world.

We have seen in the case of photography and similar processes that it is possible to store a message in the form of a permanent alteration of certain storage elements. In reinserting this information into the system, it is necessary to cause these changes to affect the messages going through the system. One of the simplest ways to do this is to have, as the storage elements which are changed, parts which normally assist in the transmission of messages, and of such a nature that the change in their character due to storage affects the manner in which they will transport messages for the entire future. In the nervous system, the neurons and the synapses are elements of this sort, and it is quite plausible that information is stored over long periods by changes in the thresholds of neurons, or, what may be regarded as another way of saying the same thing, by changes in the permeability of each synapse to messages. Many of us think, in the absence of a better explanation of the phenomenon, that the storage of information in the brain can actually occur in this way. It is conceivable for such a storage to take place either by the opening of new paths or by the closure of old ones. Apparently it is adequately established that no neurons are formed in the brain after birth. It is possible, though not certain, that no new synapses are formed, and it is a plausible conjecture that the chief changes of thresholds in the memory process are increases. If this is the case, our whole life is on the pattern of Balzac's *Peau de Chagrin*, and the very process of learning and remembering exhausts our powers of learning and remembering until life itself squanders our capital stock of power to live. It may well be that this phenomenon does occur. This is a possible explanation for a sort of senescence. The real phenomenon of senescence, however, is much too complicated to be explained in this way alone.

We have already spoken of the computing machine, and consequently the brain, as a logical machine. It is by no means trivial to consider the light cast on logic by such machines, both natural and artificial.   Here the chief work is that of Turing.[1]   We have said before that the *machina ratiocinatrix* is nothing but the *calculus ratiocinator* of Leibniz with an engine in it; and just as modern mathematical logic begins with this calculus, so it is inevitable that its present engineering development should cast a new light on logic. The science of today is operational; that is, it considers every statement as essentially concerned with possible experiments or observable processes.   According to this, the study of logic must reduce to the study of the logical machine, whether nervous or mechanical, with all its non-removable limitations and imperfections.

It may be said by some readers that this reduces logic to psychology, and that the two sciences are observably and demonstrably different.   This is true in the sense that many psychological states and sequences of thought do not conform to the canons of logic. Psychology contains much that is foreign to logic, but—and this is the important fact—any logic which means anything to us can contain nothing which the human mind—and hence the human nervous system—is unable to encompass.   *All logic is limited by the limitations of the human mind when it is engaged in that activity known as logical thinking.*

For example, we devote much of mathematics to discussions involving the infinite, but these discussions and their accompanying proofs are not infinite in fact.   No admissible proof involves more than a finite number of stages.   It is true, a proof by mathematical induction *seems* to involve an infinity of stages, but this is only apparent.   In fact, it involves just the following stages:

1. P_n is a proposition involving the number n.
2. P_n has been proved for $n = 1$.
3. If P_n is true, P_{n+1} is true.
4. Therefore, P_n is true for every positive integer n.

It is true that somewhere in our logical assumptions there must be one which validates this argument.   However, this mathematical induction is a far different thing from complete induction over an infinite set.   The same thing is true of the more refined forms of mathematical induction, such as transfinite induction, which occur in certain mathematical disciplines.

Thus some very interesting situations arise, in which we may be able—with enough time and enough computational aids—to prove

[1] Turing, A. M., "On Computable Numbers with an Application to the Entscheidungsproblem," *Proceedings of the London Mathematical Society*, Ser. 2, **42**, 230–265 (1936).

every single case of a theorem P_n; but if there is no systematic way of subsuming these proofs under a single argument independent of n, such as we find in mathematical induction, it may be impossible to prove P_n for all n. This contingency is recognized in what is known as metamathematics, the discipline so brilliantly developed by Gödel and his school.

A proof represents a logical process which has come to a definitive conclusion in a finite number of stages. However, a logical machine following definite rules need never come to a conclusion. It may go on grinding through different stages without ever coming to a stop, either by describing a pattern of activity of continually increasing complexity, or by going into a repetitive process like the end of a chess game in which there is a continuing cycle of perpetual check. This occurs in the case of some of the paradoxes of Cantor and Russell. Let us consider the class of all classes which are not members of themselves. In this class a member of itself? If it is, it is certainly not a member of itself; and if it is not, it is equally certainly a member of itself. A machine to answer this question would give the successive temporary answers: "yes," "no," "yes," "no," and so on, and would never come to equilibrium.

Bertrand Russell's solution of his own paradoxes was to affix to every statement a quantity, the so-called type, which serves to distinguish between what seems to be formally the same statement, according to the character of the objects with which it concerns itself —whether these are "things," in the simplest sense, classes of "things," classes of classes of "things," etc. The method by which we resolve the paradoxes is also to attach a parameter to each statement, this parameter being the time at which it is asserted. In both cases, we introduce what we may call a parameter of uniformization, to resolve an ambiguity which is simply due to its neglect.

We thus see that the logic of the machine resembles human logic, and, following Turing, we may employ it to throw light on human logic. Has the machine a more eminently human characteristic as well—the ability to learn? To see that it may well have even this property, let us consider two closely related notions: that of the association of ideas and that of the conditioned reflex.

In the British empirical school of philosophy, from Locke to Hume, the content of the mind was considered to be made up of certain entities known to Locke as ideas, and to the later authors as ideas and impressions. The simple ideas or impressions were supposed to exist in a purely passive mind, as free from influence on the ideas it contained as a clean blackboard is on the symbols which may be written on it.   By some sort of inner activity, hardly worthy to be called a force, these ideas were supposed to unite themselves into bundles, according to the principles of similarity, contiguity, and cause and effect.   Of these principles, perhaps the most significant was contiguity: ideas or impressions which had often occurred together in time or in space were supposed to have acquired the ability of evoking one another, so that the presence of any one of them would produce the entire bundle.

In all this there is a dynamics implied, but the idea of a dynamics had not yet filtered through from physics to the biological and psychological sciences.   The typical biologist of the eighteenth century was Linnaeus, the collector and classifier, with a point of view quite opposed to that of the evolutionists, the physiologists, the geneticists, the experimental embryologists of the present day. Indeed, with so much of the world to explore, the state of mind of the biologists could hardly have been different.   Similarly, in psychology, the notion of mental content dominated that of mental process.   This may well have been a survival of the scholastic emphasis on substances, in a world in which the noun was hypostasized and the verb carried little or no weight.   Nevertheless, the step from these static ideas to the more dynamic point of view of the present day, as exemplified in the work of Pavlov, is perfectly clear.

Pavlov worked much more with animals than with men, and he reported visible actions rather than introspective states of mind. He found in dogs that the presence of food causes the increased secretion of saliva and of gastric juice.   If then a certain visual object is shown to dogs in the presence of food and only in the presence of food, the sight of this object in the absence of food will acquire the property of being by itself able to stimulate the flow of saliva or of gastric juice.   The union by contiguity which Locke had observed introspectively in the case of ideas now becomes a similar union of patterns of behavior.

There is one important difference, however, between the point of view of Pavlov and that of Locke, and it is precisely due to this fact that Locke considers ideas and Pavlov patterns of action.   The responses observed by Pavlov tend to carry a process to a successful conclusion or to avoid a catastrophe.   Salivation is important for deglutition and for digestion, while the avoidance of what we should consider a painful stimulus tends to protect the animal from bodily injury.   Thus there enters into the conditioned reflex something that we may call *affective tone*.   We need not associate this with our own sensations of pleasure and pain, nor need we in the abstract associate it with the advantage of the animal.   The essential thing is this: that affective tone is arranged on some sort of scale from negative "pain" to positive "pleasure"; that for a considerable time, or permanently, an increase in affective tone favors all processes in the nervous system that are under way at the time and gives them a secondary power to increase affective tone; and that a decrease in affective tone tends to inhibit all processes under way at the time and gives them a secondary ability to decrease affective tone.

Biologically speaking, of course, a greater affective tone must occur predominantly in situations favorable for the perpetuation of the race, if not the individual, and a smaller affective tone in situations which are unfavorable for this perpetuation, if not disastrous. Any race not conforming to this requirement will go the way of Lewis Carroll's Bread-and-Butter Fly, and always die.   Nevertheless, even a doomed race may show a mechanism valid so long as the race lasts.   In other words, even the most suicidal apportioning of affective tone will produce a definite pattern of conduct.

Note that the mechanism of affective tone is itself a feedback mechanism.   It may even be given a diagram such as shown in Fig. 7.

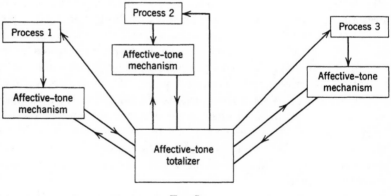

Fɪɢ. 7

Here the totalizer for affective tone combines the affective tones given by the separate affective-tone mechanisms over a short interval in the past, according to some rule which we need not specify now.   The leads back to the individual affective-tone mechanisms serve to modify the intrinsic affective tone of each process in the direction of the output of the totalizer, and this modification stands until it is modified by later messages from the totalizer.   The leads back from the totalizer to the process mechanisms serve to lower thresholds if the total affective tone is increasing, and to raise them if the total affective tone is decreasing. They likewise have a long-time effect, which endures until it is modified by another impulse from the totalizer. This lasting effect, however, is confined to those processes actually in being at the time the return message arrives, and a similar limitation also applies to the effects on the individual affective-tone mechanisms.

I wish to emphasize that I do not say that the process of the conditioned reflex operates according to the mechanism I have given; I merely say that it *could* so operate. If, however, we assume this or any simular mechanism, there are a good many things we can say concerning it. One is that this mechanism is capable of learning. It has already been recognized that the conditioned reflex is a learning mechanism, and this idea has been used in the behaviorist studies of the learning of rats in a maze. All that is needed is that the inducements or punishments used have, respectively, a positive and a negative affective tone. This is certainly the case, and the experimenter learns the nature of this affective tone by experience, not simply by *a priori* considerations.

Another point of considerable interest is that such a mechanism involves a certain set of messages which go out generally into the nervous system, to all elements which are in a state to receive them. These are the return messages from the affective-tone totalizer, and to a certain extent the messages from the affective-tone mechanisms to the totalizers. Indeed, the totalizer need not be a separate element but may merely represent some natural combinatory effect of messages arriving from the individual affective-tone mechanisms. Now, such messages "to whom it may concern" may well be sent out most efficiently, with a smallest cost in apparatus, by channels other than nervous. In a similar manner, the ordinary communication system of a mine may consist of a telephone central with the attached wiring and pieces of apparatus. When we want to empty a mine in a hurry, we do not trust to this, but break a tube of a mercaptan in the air intake. Chemical messengers like this, or like the hormones, are the simplest and most effective for a message not addressed to a specific recipient. For the moment, let me break into what I know to be pure fancy. The high emotional and consequently affective content of hormonal activity is most suggestive. This does not mean that a purely nervous mechanism is not capable of affective tone and of learning, but it does mean that in the study of this aspect of our mental activity, we cannot afford to be blind to the possibilities of hormonal transmission. It may be excessively fanciful to attach this notion to the fact that in the theories of Freud the memory—the storage function of the nervous system—and the activities of sex are both involved. Sex, on the one hand, and all affective content, on the other, contain a very strong hormonal element. This suggestion of the importance of sex and hormones has been made to me by Dr. J. Lettvin and Mr. Oliver Selfridge. While at present there is no adequate evidence to prove its validity, it is not manifestly absurd in principle.

There is nothing in the nature of the computing machine which forbids it to show conditioned reflexes. Let us remember that a computing machine in action is more than the concatenation of relays and storage mechanisms which the designer has built into it. It also contains the content of its storage mechanisms, and this content is never completely cleared in the course of a single run. We have already seen that it is the run rather than the entire existence of the mechanical structure of the computing machine which corresponds to the life of the individual. We have also seen that in the nervous computing machine it is highly probable that information is stored largely as changes in the permeability of the synapses, and it is perfectly possible to construct artificial machines where information is stored in that way. It is perfectly possible, for example, to cause any message going into storage to change in a permanent or semi-permanent way the grid bias of one or of a number of vacuum tubes, and thus to alter the numerical value of the summation of impulses which will make the tube or tubes fire.

A more detailed account of learning apparatus in computing and control machines, and the uses to which it may be put, may well be left to the engineer rather than to a preliminary book like this one. It is perhaps better to devote the rest of this chapter to the more developed, normal uses of modern computing machines. One of the chief of these is in the solution of partial differential equations. Even linear partial differential equations require the recording of an enormous mass of data to set them up, as the data involve the accurate description of functions of two or more variables. With equations of the hyperbolic type, like the wave equation, the typical problem is that of solving the equation when the initial data are given, and this can be done in a progressive manner from the initial data to the results at any given later time. This is largely true of equations of the parabolic type as well. When it comes to equations of the elliptic type, where the natural data are boundary values rather than initial values, the natural methods of solution involve an iterative process of successive approximation. This process is repeated a very large number of times, so that very fast methods, such as those of the modern computing machine, are almost indispensable.

In non-linear partial differential equations, we miss what we have in the case of the linear equations—a reasonably adequate, purely mathematical theory. Here computational methods are not only important for the handling of particular numerical cases, but, as von Neumann has pointed out, we need them in order to form that acquaintance with a large number of particular cases without which we can scarcely formulate a general theory. To some extent this has been done with the aid of very expensive experimental apparatus, such as wind tunnels. It is in this way that we have become acquainted with the more complicated properties of shock waves, slip surfaces, turbulence, and the like, for which we are scarcely in a position to give an adequate mathematical theory. How many undiscovered phenomena of similar nature there may be, we do not know. The analogue machines are so much less accurate, and in many cases so much slower, than the digital machines that the latter give us much more promise for the future.

It is already becoming clear in the use of these new machines that they demand purely mathematical techniques of their own, quite different from those in use in manual computation or in the use of machines of smaller capacity. For example, even the use of machines for computing determinants of moderately high order or for the simultaneous solution of twenty or thirty simultaneous linear equations shows difficulties which do not arise when we study analogous problems of small order. Unless care is exercised in setting up a problem, these may completely deprive the solution of any significant figures whatever. It is a commonplace to say that fine, effective tools like the ultra-rapid computing machine are out of place in the hands of those not possessing a sufficient degree of technical skill to take full advantage of them. The ultra-rapid computing machine will certainly not decrease the need for mathematicians with a high level of understanding and technical training.

In the mechanical or electrical construction of computing machines, there are a few maxims which deserve consideration. One is that mechanisms which are relatively frequently used, such as multiplying or adding mechanisms, should be in the form of relatively standardized assemblages adapted for one particular use and no other, while those of more occasional use should be assembled for the moment of use out of elements also available for other purposes. Closely related to this consideration is the one that in these more general mechanisms the component parts should be available in accordance with their general properties, and should not be allotted permanently to a specific association with other pieces of apparatus. There should be some part of the apparatus, like an automatic telephone-switching exchange, which will search for free components and connectors of the various sorts and allot them as they are needed. This will eliminate much of the very large expense which is due to having a great number of unused elements which cannot be used unless their entire large assembly is used. We shall find this principle is very important when we come to consider traffic problems and overloading in the nervous system.

As a final remark, let me point out that a large computing machine, whether in the form of mechanical or electric apparatus or in the form of the brain itself, uses up a considerable amount of power, all of which is wasted and dissipated in heat. The blood leaving the brain is a fraction of a degree warmer than that entering it. No other computing machine approaches the economy of energy of the brain. In a large apparatus like the Eniac or Edvac, the filaments of the tubes consume a quantity of energy which may well be measured in kilowatts, and unless adequate ventilating and cooling apparatus is provided, the system will suffer from what is the mechanical equivalent of pyrexia, until the constants of the machine are radically changed by the heat, and its performance breaks down. Nevertheless, the energy spent per individual operation is almost vanishingly small, and does not even begin to form an adequate measure of the performance of the apparatus. The mechanical brain does not secrete thought "as the liver does bile," as the earlier materialists claimed, nor does it put it out in the form of energy, as the muscle puts out its activity. Information is information, not matter or energy. No materialism which does not admit this can survive at the present day.

VI

Gestalt and Universals

Among other things which we have discussed in the previous chapter is the possibility of assigning a neural mechanism to Locke's theory of the association of ideas. According to Locke, this occurs according to three principles: the principle of contiguity, the principle of similarity, and the principle of cause and effect. The third of these is reduced by Locke, and even more definitively by Hume, to nothing more than constant concomitance, and so is subsumed under the first, that of contiguity. The second, that of similarity, deserves a more detailed discussion.

How do we recognize the identity of the features of a man, whether we see him in profile, in three-quarters face, or in full face? How do we recognize a circle as a circle, whether it is large or small, near or far; whether, in fact, it is in a plane perpendicular to a line from the eye meeting it in the middle, and is seen as a circle, or has some other orientation, and is seen as an ellipse? How do we see faces and animals and maps in clouds, or in the blots of a Rorschach test? All these examples refer to the eye, but similar problems extend to the other senses, and some of them have to do with intersensory relations. How do we put into words the call of a bird or the stridulations of an insect? How do we identify the roundness of a coin by touch?

For the present, let us confine ourselves to the sense of vision. One important factor in the comparison of form of different objects is certainly the interaction of the eye and the muscles, whether they are the muscles within the eyeball, the muscles moving the eyeball, the muscles moving the head, or the muscles moving the body as a whole. Indeed, some form of this visual-muscular feedback system is important as low in the animal kingdom as the flatworms. There the negative phototropism, the tendency to avoid the light, seems to be controlled by the balance of the impulses from the two eyespots. This balance is fed back to the muscles of the trunk, turning the body away from the light, and, in combination with the general impulse to move forward, carries the animal into the darkest region accessible. It is interesting to note that a combination of a pair of photocells with appropriate amplifiers, a Wheatstone bridge for balancing their outputs, and further amplifiers controlling the input into the two motors of a twinscrew mechanism would give us a very adequate negatively phototropic control for a little boat. It would be difficult or impossible for us to compress this mechanism into the dimensions that a flatworm can carry; but here we merely have another exemplification of the fact that must by now be familiar to the reader, that living mechanisms tend to have a much smaller space scale than the mechanisms best suited to the techniques of human artificers, although, on the other hand, the use of electrical techniques gives the artificial mechanism an enormous advantage in speed over the living organism.

Without going through all the intermediate stages, let us come at once to the eye-muscle feedbacks in man. Some of these are of purely homeostatic nature, as when the pupil opens in the dark and closes in the light, thus tending to confine the flow of light into the eye between narrower bounds than would otherwise be possible. Others concern the fact that the human eye has economically confined its best form and color vision to a relatively small fovea, while its perception of motion is better on the periphery. When the peripheral vision has picked up some object conspicuous by brilliancy or light contrast or color or above all by motion, there is a reflex feedback to bring it into the fovea. This feedback is accompanied by a complicated system of interlinked subordinate feedbacks, which tend to converge the two eyes so that the object attracting attention is in the same part of the visual field of each, and to focus the lens so that its outlines are as sharp as possible. These actions are supplemented by motions of the head and body, by which we bring the object into the center of vision if this cannot be done readily by a motion of the eyes alone, or by which we bring an object outside the visual field picked up by some other sense into that field. In the case of objects with which we are more familiar in one angular orientation than another—writing, human faces, landscapes, and the like—there is also a mechanism by which we tend to pull them into the proper orientation.

All these processes can be summed up in one sentence: we tend to bring any object that attracts our attention into a standard position and orientation, so that the visual image which we form of it varies within as small a range as possible. This does not exhaust the processes which are involved in perceiving the form and meaning of the object, but it certainly facilitates all later processes tending to this end. These later processes occur in the eye and in the visual cortex. There is considerable evidence that for a considerable number of stages each step in this process diminishes the number of neuron channels involved in the transmission of visual information, and brings this information one step nearer to the form in which it is used and is preserved in the memory.

The first step in this concentration of visual information occurs in the transition between the retina and the optic nerve. It will be noted that while in the fovea there is almost a one-one correspondence between the rods and cones and the fibers of the optic nerve, the correspondence on the periphery is such that one optic nerve fiber corresponds to ten or more end organs. This is quite understandable, in view of the fact that the chief function of the peripheral fibers is not so much vision itself as a pickup for the centering and focusing-directing mechanism of the eye.

One of the most remarkable phenomena of vision is our ability to recognize an outline drawing. Clearly, an outline drawing of, say, the face of a man, has very little resemblance to the face itself in color, or in the massing of light and shade, yet it may be a most recognizable portrait of its subject. The most plausible explanation of this is that, somewhere in the visual process, outlines are emphasized and some other aspects of an image are minimized in importance. The beginning of these processes is in the eye itself. Like all senses, the retina is subject to accommodation; that is, the constant maintenance of a stimulus reduces its ability to receive and to transmit that stimulus. This is most markedly so for the receptors which record the interior of a large block of images with constant color and illumination, for even the slight fluctuations of focus and point of fixation which are inevitable in vision do not change the character of the image received. It is quite different on the boundary of two contrasting regions. Here these fluctuations produce an alternation between one stimulus and another, and this alternation, as we see in the phenomenon of after-images, not only does not tend to exhaust the visual mechanism by accommodation but even tends to enhance its sensitivity. This is true whether the contrast between the two adjacent regions is one of light intensity or of color. As a comment on these facts, let us note that three-quarters of the fibers in the optic nerve respond only to the flashing "on" of illumination. We thus find that the eye receives its most intense impression at boundaries, and that every visual image in fact has something of the nature of a line drawing.

Probably not all of this action is peripheral. In photography, it is known that certain treatments of a plate increase its contrasts, and such phenomena, which are of non-linearity, are certainly not beyond what the nervous system can do. They are allied to the phenomena of the telegraph-type repeater, which we have already mentioned. Like this, they use an impression which has not been blurred beyond a certain point to trigger a new impression of a standard sharpness. At any rate, they decrease the total unusable information carried by an image, and are probably correlated with a part of the reduction of the number of transmission fibers found at various stages of the visual cortex.

We have thus designated several actual or possible stages of the diagrammatization of our visual impressions. We center our images around the focus of attention and reduce them more or less to outlines. We have now to compare them with one another, or at any rate with a standard impression stored in memory, such as "circle" or "square." This may be done in several ways. We have given a rough sketch which indicates how the Lockean principle of contiguity in association may be mechanized. Let us notice that the principle of contiguity also covers much of the other Lockean principle of similarity. The different aspects of the same object are often to be seen in those processes which bring it to the focus of attention, and of other motions which lead us to see it, now at one distance and now at another, now from one angle and now from a distinct one. This is a general principle, not confined in its application to any particular sense and doubtless of much importance in the comparison of our more complicated experiences. It is nevertheless probably not the only process which leads to the formation of our specifically visual general ideas, or, as Locke would call them, "complex ideas." The structure of our visual cortex is too highly organized, too specific, to lead us to suppose that it operates by what is after all a highly generalized mechanism. It leaves us the impression that we are here dealing with a special mechanism which is not merely a temporary assemblage of general-purpose elements with interchangeable parts, but a permanent sub-assembly like the adding and multiplying assemblies of a computing machine. Under the circumstances, it is worth considering how such a sub-assembly might possibly work and how we should go about designing it.

The possible perspective transformations of an object form what is known as a group, in the sense in which we have already defined one in Chapter II. This group defines several sub-groups of transformations: the affine group, in which we consider only those transformations which leave the region at infinity untouched; the homogeneous dilations about a given point, in which one point, the directions of the axes, and the equality of scale in all directions are preserved; the transformations preserving length; the rotations in two or three dimensions about a point; the set of all translations; and so on. Among these groups, the ones we have just mentioned are continuous; that is, the operations belonging to them are determined by the values of a number of continuously varying parameters in an appropriate space. They thus form multi-dimensional configurations in n-space, and contain sub-sets of transformations which constitute regions in such a space.

Now, just as a region in the ordinary two-dimensional plane is covered by the process of scanning known to the television engineer, by which a nearly uniformly distributed set of sample positions in that region is taken to represent the whole, so every region in a group-space, including the whole of such a space, can be represented by a process of *group scanning*. In such a process, which is by no means confined to a space of three dimensions, a net of positions in the space is traversed in a one-dimensional sequence, and this net of positions is so distributed that it comes near to every position in the region, in some appropriately defined sense. It will thus contain positions as near to any we wish as may be desired. If these "positions," or sets of parameters, are actually used to generate the appropriate transformations, it means that the results of transforming a given figure by these transformations will come as near as we wish to any given transformation of the figure by a transformation operator lying in the region desired. If our scanning is fine enough, and the region transformed has the maximum dimensionality of the regions transformed by the group considered, this means that the transformations actually traversed will give a resulting region overlapping *any* transform of the original region by an amount which is as large a fraction of its area as we wish.

Let us then start with a fixed comparison region and a region to be compared with it. If at any stage of the scanning of the group of transformations the image of the region to be compared under some one of the transformations scanned coincides more perfectly with the fixed pattern than a given tolerance allows, this is recorded, and the two regions are said to be alike. If this happens at no stage of the scanning process, they are said to be unlike. This process is perfectly adapted to mechanization, and serves as a method to identify the shape of a figure independently of its size or its orientation or of whatever transformations may be included in the group-region to be scanned.

If this region is not the entire group, it may well be that region A seems like region B, and that region B seems like region C, while region A does not seem like region C. This certainly happens in reality. A figure may not show any particular resemblance to the same figure inverted, at least in so far as the immediate impression— one not involving any of the higher processes—is concerned. Nevertheless, at each stage of its inversion, there may be a considerable range of neighboring positions which appear similar. The universal "ideas" thus formed are not perfectly distinct but shade into one another.

There are other more sophisticated means of using group scanning to abstract from the transformations of a group. The groups which we here consider have a "group measure," a probability density which depends on the transformation group itself and does not change when all the transformations of the group are altered by being preceded or followed by any specific transformation of the group. It is possible to scan the group in such a way that the density of scanning of any region of a considerable class—that is, the amount of time which the variable scanning element passes within the region in any complete scanning of the group—is closely proportional to its group measure. In the case of such a uniform scanning, if we have any quantity depending on a set S of elements transformed by the group, and if this set of elements is transformed by all the transformations of the group, let us designate the quantity depending on S by $Q(S)$, and let us use TS to express the transform of the set S by the transformation T of the group. Then $Q(TS)$ will be the value of the quantity replacing $Q(S)$ when S is replaced by TS. If we average or integrate this with respect to the group measure for the group of transformations T, we shall obtain a quantity which we may write in some such form as

$$\int Q(TS)\, dT \qquad\qquad (6.01)$$

where the integration is over the group measure. Quantity 6.01 will be identical for all sets S interchangeable with one another under the transformations of the group, that is, for all sets S which have in some sense the same form or *Gestalt*. It is possible to obtain an approximate comparability of form where the integration in Quantity 6.01 is over less than the whole group, if the integrand $Q(TS)$ is small over the region omitted. So much for group measure.

In recent years, there has been a good deal of attention to the problem of the prosthesis of one lost sense by another. The most dramatic of the attempts to accomplish this has been the design of reading devices for the blind, to work by the use of photoelectric cells. We shall suppose that these efforts are confined to printed matter, and even to a single type face or to a small number of type faces. We shall also suppose that the alignment of the page, the centering of the lines, the traverse from line to line are taken care of either manually or, as they may well be, automatically. These processes correspond, as we may see, to the part of our visual *Gestalt* determination which depends on muscular feedbacks and the use of our normal centering, orienting, focusing, and converging apparatus. There now ensues the problem of determining the shapes of the individual letters as the scanning apparatus passes over them in sequence. It has been suggested that this be done by the use of several photoelectric cells placed in a vertical sequence, each attached to a sound-making apparatus of a different pitch. This can be done with the black of the letters registering either as silence or as sound. Let us assume the latter case, and let us assume three photocell receptors above one another. Let them record as the three notes of a chord, let us say, with the highest note on top and the lowest note below. Then the letter capital F, let us say, will record

————————————   Duration of upper note

————————   Duration of middle note

——   Duration of lower note

The letter capital Z will record

————————————

——

————————————

the letter capital O

——

——   ——

——

and so on.   With the ordinary help given by our ability to interpret, it should not be too difficult to read such an auditory code, not more difficult than to read Braille, for instance.

However, all this depends on one thing: the proper relation of the photocells to the vertical height of the letters.   Even with standardized type faces, there still are great variations in the size of the type. Thus it is desirable for us to be able to pull the vertical scale of the scanning up or down, in order to reduce the impression of a given letter to a standard.   We must at least have at our disposal, manually or automatically, some of the transformations of the vertical dilation group.

There are several ways we might do this.   We might allow for a mechanical vertical adjustment of our photocells.   On the other hand, we might use a rather large vertical array of photocells and change the pitch assignment with the size of type, leaving those above and below the type silent.   This may be done, for example, with the aid of a schema of two sets of connectors, the inputs coming up from the photocells, and leading to a series of switches of wider and wider divergence, and the outputs a series of vertical lines, as in

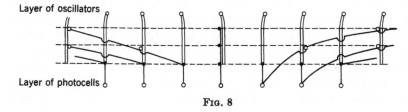

Fɪɢ. 8

Fig. 8.   Here the single lines represent the leads from the photocells, the double lines the leads to the oscillators, the circles on the dotted lines the points of connections between incoming and outgoing leads, and the dotted lines themselves the leads whereby one or another of a bank of oscillators is put into action.   This was the device, to which we have referred in the introduction, designed by McCulloch for the purpose of adjusting to the height of the type face.   In the first design, the selection between dotted line and dotted line was manual.

This was the figure which, when shown to Dr. von Bonin, suggested the fourth layer of the visual cortex.   It was the connecting circles which suggested the neuron cell bodies of this layer, arranged in sub-layers of uniformly changing horizontal density, and size changing in the opposite direction to the density.   The horizontal leads are probably fired in some cyclical order.   The whole apparatus seems quite suited to the process of group scanning.   There must of course be some process of recombination in time of the upper outputs.

This then was the device suggested by McCulloch as that actually used in the brain in the detection of visual *Gestalt*. It represents a type of device usable for any sort of group scanning. Something similar occurs in other senses as well. In the ear, the transposition of music from one fundamental pitch to another is nothing but a translation of the logarithm of the frequency, and may consequently be performed by a group-scanning apparatus.

A group-scanning assembly thus has well-defined, appropriate anatomical structure. The necessary switching may be performed by independent horizontal leads which furnish enough stimulation to shift the thresholds in each level to just the proper amount to make them fire when the lead comes on. While we do not know all the details of the performance of the machinery, it is not at all difficult to conjecture a possible machine conforming to the anatomy. In short, the group-scanning assembly is well adapted to form the sort of permanent sub-assembly of the brain corresponding to the adders or multipliers of the numerical computing machine.

Lastly, the scanning apparatus should have a certain intrinsic period of operation which should be identifiable in the performance of the brain. The order of magnitude of this period should show in the minimum time required for making direct comparison of the shapes of objects different in size. This can be done only when the comparison is between two objects not too different in size; otherwise, it is a long-time process, suggestive of the action of a non-specific assembly. When direct comparison seems to be possible, it appears to take a time of the order of magnitude of a tenth of a second. This also seems to accord with the order of magnitude of the time needed by excitation to stimulate all the layers of transverse connectors in cyclical sequence.

While this cyclical process then might be a locally determined one, there is evidence that there is a widespread synchronism in different parts of the cortex, suggesting that it is driven from some clocking center. In fact, it has the order of frequency appropriate for the alpha rhythm of the brain, as shown in electroencephalograms. We may suspect that this alpha rhythm is associated with form perception, and that it partakes of the nature of a sweep rhythm, like the rhythm shown in the scanning process of a television apparatus. It disappears in deep sleep, and seems to be obscured and overlaid with other rhythms, precisely as we might expect, when we are actually looking at something and the sweep rhythm is acting as something like a carrier for other rhythms and activities. It is most marked when the eyes are closed in waking, or when we are staring into space at nothing in particular, as in the condition of abstraction of a yogi,[1] when it shows an almost perfect periodicity.

We have just seen that the problem of sensory prosthesis—the problem of replacing the information normally conveyed through a lost sense by information through another sense still available—is important and not necessarily insoluble. What makes it more hopeful is the fact that the memory and association areas, normally approached through one sense, are not locks with a single key but are available to store impressions gathered from other senses than the one to which they normally belong. A blinded man, as distinguished perhaps from one congenitally blind, not only retains visual memories earlier in date than his accident but is even able to store tactile and auditory impressions in a visual form. He may feel his way around a room, and yet have an image of how it ought to look.

Thus a part of his normal visual mechanism is accessible to him. On the other hand, he has lost more than his eyes: he has also lost the use of that part of his visual cortex which may be regarded as a fixed assembly for organizing the impressions of sight. It is necessary to equip him not only with artificial visual receptors but with an artificial visual cortex, which will translate the light impressions on his new receptors into a form so related to the normal output of his visual cortex that objects which ordinarily look alike will now sound alike.

Thus the criterion of the possibility of such a replacement of sight by hearing is at least in part a comparison between the number of recognizably different visual patterns and recognizably different auditory patterns *at the cortical level*. This is a comparison of amounts of information. In view of the somewhat similar organization of the different parts of the sensory cortex, it will probably not differ very much from a comparison between the areas of the two parts of the cortex. This is about 100:1 as between sight and sound. If all the auditory cortex were used for vision, we might expect to get a quantity of reception of information about 1 per cent of that coming in through the eye. On the other hand, our usual scale for the estimation of vision is in terms of the relative distance at which a certain degree of resolution of pattern is obtained, and thus a 10/100 vision means an amount of flow of information about 1 per

[1] Personal communication of Dr. W. Grey Walter, of Bristol, England.

cent of normal. This is very poor vision; it is, however, definitely not blindness, nor do people with this amount of vision necessarily consider themselves as blind.

In the other direction, the picture is even more favorable. The eye can detect all of the nuances of the ear with the use of only 1 per cent of its facilities, and still leave a vision of about 95/100, which is substantially perfect. Thus the problem of sensory prosthesis is an extremely hopeful field of work.

VII

Cybernetics and Psychopathology

It is necessary that I commence this chapter with a disavowal. On the one hand, I am not a psychopathologist nor a psychiatrist, and lack any experience in a field where the guidance of experience is the only trustworthy one. On the other hand, our knowledge of the normal performance of the brain and the nervous system, and *a fortiori* our knowledge of their abnormal performance, is far from having reached that state of perfection where an *a priori* theory can command any confidence. I therefore wish to disclaim in advance any assertion that any particular entity in psychopathology, as for example any of the morbid conditions described by Kraepelin and his disciples, is due to a specific type of defect in the organization of the brain as a computing machine. Those who may draw such specific conclusions from the considerations of this book do so at their own risk.

Nevertheless, the realization that the brain and the computing machine have much in common may suggest new and valid approaches to psychopathology and even to psychiatrics. These begin with perhaps the simplest question of all: how the brain avoids gross blunders, gross miscarriages of activity, due to the malfunction of individual components. Similar questions referring to the computing machine are of great practical importance, for here a chain of operations, each covering a fraction of a millisecond, may last a matter of hours or days. It is quite possible for a chain of computational operations to involve 10^9 separate steps. Under these circumstances, the chance that at least one operation will go amiss is very far from negligible, even though, it is true, the reliability

144

of modern electronic apparatus has far exceeded the most sanguine expectations.

In ordinary computational practice by hand or by desk machines, it is the custom to check every step of the computation and, when an error is found, to localize it by a backward process starting from the first point where the error is noted. To do this with a high-speed machine, the check must proceed with the speed of the original machine, or the whole effective order of speed of the machine will conform to that of the slower process of checking. Furthermore, if the machine is made to keep all intermediate records of its performance, its complication and bulk will be increased to an intolerable point, by a factor which is likely to be enormously greater than 2 or 3.

A much better method of checking, and in fact the one generally used in practice, is to refer every operation simultaneously to two or three separate mechanisms. In the case of the use of two such mechanisms, their answers are automatically collated against each other; and if there is a discrepancy, all data are transferred to permanent storage, the machine stops, and a signal is sent to the operator that something is wrong. The operator then compares the results, and is guided by them in his search for the malfunctioning part, perhaps a tube which has burnt out and needs replacement. If three separate mechanisms are used for each stage and single misfunctions are as rare as they are in fact, there will practically always be an agreement between two of the three mechanisms, and this agreement will give the required result. In this case, the collation mechanism accepts the majority report, and the machine need not stop; but there is a signal indicating where and how the minority report differs from the majority report. If this occurs at the first moment of discrepancy, the indication of the position of the error may be very precise. In a well-designed machine, no particular element is assigned to a particular stage in the sequence of operations, but at each stage there is a searching process, quite similar to that used in automatic telephone exchanges, which finds the first available element of a given sort and switches it into the sequence of operations. In this case, the removal and replacement of defective elements need not be the source of any appreciable delay.

It is conceivable and not implausible that at least two of the elements of this process are also represented in the nervous system. We can hardly expect that any important message is entrusted for transmission to a single neuron, nor that any important operation is entrusted to a single neuronal mechanism. Like the computing machine, the brain probably works on a variant of the famous principle expounded by Lewis Carroll in *The Hunting of the Snark*: "What I tell you three times is true."   It is also improbable that the various channels available for the transfer of information generally go from one end of their course to the other without anastomosing. It is much more probable that when a message comes in to a certain level of the nervous system, it may leave that point and proceed to the next by one or more alternative members of what is known as an "internuncial pool."   There may be parts of the nervous system, indeed, where this interchangeability is much limited or abolished, and these are likely to be such highly specialized parts of the cortex as those which serve as the inward extensions of the organs of special sense.   Still, the principle holds, and probably holds most clearly for the relatively unspecialized cortical areas which serve the purpose of association and of what we call the higher mental functions.

So far we have been considering errors in performance which are normal, and pathological only in an extended sense.   Let us now turn to those which are much more clearly pathological.   Psychopathology has been rather a disappointment to the instinctive materialism of the doctors, who have taken the point of view that every disorder must be accompanied by material lesions of some specific tissue involved.   It is true that specific brain lesions, such as injuries, tumors, clots, and the like, may be accompanied by psychic symptoms, and that certain mental diseases, such as paresis, are the sequellae of general bodily disease and show a pathological condition of the brain tissue; but there is no way of identifying the brain of a schizophrenic of one of the strict Kraepelin types, nor of a manic-depressive patient, nor of a paranoiac.   These disorders we call *functional*, and this distinction seems to contravene the dogma of modern materialism that every disorder in function has some physiological or anatomical basis in the tissues concerned.

This distinction between functional and organic disorders receives a great deal of light from the consideration of the computing machine. As we have already seen, it is not the empty physical structure of the computing machine that corresponds to the brain—to the adult brain, at least—but the combination of this structure with the instructions given it at the beginning of a chain of operations and with all the additional information stored and gained from outside in the course of this chain.   This information is stored in some physical form—in the form of memory—but part of it is in the form of circulating memories, with a physical basis which vanishes when the machine is shut down or the brain dies, and part in the form of long-time memories, which are stored in a way at which we can only guess, but probably also in a form with a physical basis which vanishes at death. There is no way yet known for us to recognize in the cadaver what the threshold of a given synapse has been in life; and even if we knew this, there is no way we can trace out the chain of neurons and synapses communicating with this, and determine the significance of this chain for the ideational content which it records.

There is therefore nothing surprising in considering the functional mental disorders as fundamentally diseases of memory, of the circulating information kept by the brain in the active state, and of the long-time permeability of synapses. Even the grosser disorders such as paresis may produce a large part of their effects not so much by the destruction of tissue which they involve and the alteration of synaptic thresholds as by the secondary disturbances of traffic—the overload of what remains of the nervous system and the re-routing of messages—which must follow such primary injuries.

In a system containing a large number of neurons, circular processes can hardly be stable for long periods of time. Either, as in the case of memories belonging to the specious present, they run their course, dissipate themselves, and die out, or they comprehend more and more neurons in their system, until they occupy an inordinate part of the neuron pool. This is what we should expect to be the case in the malignant worry which accompanies anxiety neuroses. In such a case, it is possible that the patient simply does not have the room, the sufficient number of neurons, to carry out his normal processes of thought. Under such conditions, there may be less going on in the brain to load up the neurons not yet affected, so that they are all the more readily involved in the expanding process. Furthermore, the permanent memory becomes more and more deeply involved, and the pathological process which occurred at first at the level of the circulating memories may repeat itself in a more intractable form at the level of the permanent memories. Thus what started as a relatively trivial and accidental reversal of stability may build itself up into a process totally destructive to the ordinary mental life.

Pathological processes of a somewhat similar nature are not unknown in the case of mechanical or electrical computing machines. A tooth of a wheel may slip under just such conditions that no tooth with which it engages can pull it back into its normal relations, or a high-speed electrical computing machine may go into a circular process which there seems to be no way to stop. These contingencies may depend on a highly improbable instantaneous configuration of the system, and, when remedied, may never—or very rarely—repeat themselves. However, when they occur, they temporarily put the machine out of action.

How do we deal with these accidents in the use of the machine? The first thing which we try is to clear the machine of all information, in the hope that when it starts again with different data the difficulty may not recur. Failing this, if the difficulty is in some point permanently or temporarily inaccessible to the clearing mechanism, we shake the machine or, if it is electrical, subject it to an abnormally large electrical impulse, in the hope that we may reach the inaccessible part and throw it into a position where the false cycle of its activities will be interrupted. If even this fails, we may disconnect an erring part of the apparatus, for it is possible that what yet remains may be adequate for our purpose.

Now there is no normal process except death which completely clears the brain from all past impressions; and after death, it is impossible to set it going again. Of all normal processes, sleep comes the nearest to a non-pathological clearing. How often we find that the best way to handle a complicated worry or an intellectual muddle is to sleep over it! However, sleep does not clear away the deeper memories, nor indeed is a sufficiently malignant state of worry compatible with an adequate sleep. We are thus often forced to resort to more violent types of intervention in the memory cycle. The more violent of these involve a surgical intervention into the brain, leaving behind it permanent damage, mutilation, and the abridgment of the powers of the victim, as the mammalian central nervous system seems to possess no powers whatever of regeneration. The principal type of surgical intervention which has been practiced is known as prefrontal lobotomy, and consists in the removal or isolation of a portion of the prefrontal lobe of the cortex. It has recently been having a certain vogue, probably not unconnected with the fact that it makes the custodial care of many patients easier. Let me remark in passing that killing them makes their custodial care still easier. However, prefrontal lobotomy does seem to have a genuine effect on malignant worry, not by bringing the patient nearer to a solution of his problems but by damaging or destroying the capacity for maintained worry, known in the terminology of another profession as the *conscience*. More generally, it appears to limit all aspects of the circulating memory, the ability to keep in mind a situation not actually presented.

The various forms of shock treatment—electric, insulin, metrazol—are less drastic methods of doing a very similar thing. They do not destroy brain tissue or at least are not intended to destroy it, but they do have a decidedly damaging effect on the memory.   In so far as this concerns the circulating memory, and in so far as this memory is chiefly damaged for the recent period of mental disorder, and is probably scarcely worth preserving anyhow, shock treatment has something definite to recommend it as against lobotomy; but it is not always free from deleterious effects on the permanent memory and the personality.   As it stands at present, it is another violent, imperfectly understood, imperfectly controlled method to interrupt a mental vicious circle.   This does not prevent its being in many cases the best thing we can do at present.

Lobotomy and shock treatment are methods which by their very nature are more suited to handle vicious circulating memories and malignant worries than the deeper-seated permanent memories, though it is not impossible that they may have some effect here too. As we have said, in long-established cases of mental disorder, the permanent memory is as badly deranged as the circulating memory. We do not seem to possess any purely pharmaceutical or surgical weapon for intervening differentially in the permanent memory. This is where psychoanalysis and other similar psychotherapeutic measures come in.   Whether psychoanalysis is taken in the orthodox Freudian sense or in the modified senses of Jung and of Adler, or whether our psychotherapy is not strictly psychoanalytic at all, our treatment is clearly based on the concept that the stored information of the mind lies on many levels of accessibility and is much richer and more varied than that which is accessible by direct unaided introspection; that it is vitally conditioned by affective experiences which we cannot always uncover by such introspection, either because they never were made explicit in our adult language, or because they have been buried by a definite mechanism, affective though generally involuntary; and that the content of these stored experiences, as well as their affective tone, conditions much of our later activity in ways which may well be pathological.   The technique of the psychoanalyst consists in a series of means to discover and interpret these hidden memories, to make the patient accept them for what they are and by their acceptance modify, if not their content, at least the affective tone they carry, and thus make them less harmful.   All this is perfectly consistent with the point of view of this book.   It perhaps explains, too, why there are circumstances where a joint use of shock treatment and psychotherapy is indicated, combining a physical or pharmacological therapy for the phenomena of reverberation in the nervous system, and a psychological therapy for the long-time memories which, without interference, might re-establish from within the vicious circle broken up by the shock treatment.

We have already mentioned the traffic problem of the nervous system. It has been commented on by many writers, such as D'Arcy Thompson,[1] that each form of organization has an upper limit of size, beyond which it will not function. Thus the insect organization is limited by the length of tubing over which the spiracle method of bringing air by diffusion directly to the breathing tissues will function; a land animal cannot be so big that the legs or other portions in contact with the ground will be crushed by its weight; a tree is limited by the mechanism for transferring water and minerals from the roots to the leaves, and the products of photosynthesis from the leaves to the roots; and so on. The same sort of thing is observed in engineering constructions. Skyscrapers are limited in size by the fact that when they exceed a certain height, the elevator space needed for the upper stories consumes an excessive part of the cross section of the lower floors. Beyond a certain span, the best-possible suspension bridge which can be built out of materials with given elastic properties will collapse under its own weight; and beyond a certain greater span, *any* structure built of a given material or materials will collapse under its own weight. Similarly, the size of a single telephone central, built according to a constant, non-expanding plan, is limited, and this limitation has been very thoroughly studied by telephone engineers.

In a telephone system, the important limiting factor is the fraction of the time during which a subscriber will find it impossible to put a call through. A 99 per cent chance of success will certainly be satisfactory for even the most exacting; 90 per cent of successful calls is probably good enough to permit business to be carried on with reasonable facility. A success of 75 per cent is already annoying but will permit business to be carried on after a fashion; while if half the calls end in failures, subscribers will begin to ask to have their telephones taken out. Now, these represent over-all figures. If the calls go through n distinct stages of switching, and probability of failure is independent and equal for each stage, in order to get a probability of total success equal to p, the probability of success at each stage must be $p^{1/n}$. Thus to obtain a 75 per cent chance of the completion of the call after five stages, we must have about 95 per cent chance of success per stage. To obtain a 90 per cent perform-

---

[1] Thompson, D'Arcy, *On Growth and Form*, Amer. ed., The Macmillan Company, New York, 1942.

ance, we must have 98 per cent chance of success at each stage.   To obtain a 50 per cent performance, we must have 87 per cent chance of success at each stage.   It will be seen that the more stages which are involved, the more rapidly the service becomes extremely bad when a critical level of failure for the individual call is exceeded, and extremely good when this critical level of failure is not quite reached. Thus a switching service involving many stages and designed for a certain level of failure shows no obvious signs of failure until the traffic comes up to the edge of the critical point, when it goes completely to pieces, and we have a catastrophic traffic jam.

Man, with the best-developed nervous system of all the animals, with behavior that probably depends on the longest chains of effectively operated neuronic chains, is then likely to perform a complicated type of behavior efficiently very close to the edge of an overload, when he will give way in a serious and catastrophic way. This overload may take place in several ways: either by an excess in the amount of traffic to be carried, by a physical removal of channels for the carrying of traffic, or by the excessive occupation of such channels by undesirable systems of traffic, like circulating memories which have increased to the extent of becoming pathological worries. In all these cases, a point will come—quite suddenly—when the normal traffic will not have space enough allotted to it, and we shall have a form of mental breakdown, very possibly amounting to insanity.

This will first affect the faculties or operations involving the longest chains of neurons.   There is appreciable evidence that these are precisely the processes which are recognized to be the highest in our ordinary scale of valuation.   The evidence is this: a rise in temperature within nearly physiological limits is known to produce an increase in the ease of performance of most if not of all neuronic processes.   This is greater for the higher processes, roughly in the order of our usual estimate of their degree of "highness."   Now, any facilitation of a process in a single neuron-synapse system should be cumulative as the neuron is combined in series with other neurons. Thus the amount of assistance a process receives through a rise in temperature is a rough measure of the length of the neuron chain it involves.

We thus see that the superiority of the human brain to others in the length of the neuron chains it employs is a reason why mental disorders are certainly most conspicuous and probably most common in man.   There is another more specific way of considering a very similar matter.   Let us first consider two brains geometrically similar, with the weights of gray and of white matter related by the same factor of proportionality, but with different linear dimensions in the ratio $A:B$. Let the volume of the cell bodies in the gray matter and the cross sections of the fibers in the white matter be of the same size in both brains. Then the number of cell bodies in the two cases bears the ratio $A^3:B^3$, and the number of long-distance connectors the ratio $A^2:B^2$. This means that for the same density of activity in the cells, the density of activity in the fibers is $A:B$ times as great in the case of the large brain as in that of the small brain.

If we compare the human brain with that of a lower mammal, we shall find that it is much more convoluted. The relative thickness of the gray matter is much the same, but it is spread over a far more involved system of gyri and sulci. The effect of this is to increase the amount of gray matter at the expense of the amount of white matter. Within a gyrus, this decrease of the white matter is largely a decrease in length rather than in number of fibers, as the opposing folds of a gyrus are nearer together than they would be on a smooth-surfaced brain of the same size. On the other hand, when it comes to the connectors between different gyri, the distance they have to run is increased if anything by the convolution of the brain. Thus the human brain would seem to be fairly efficient in the matter of the short-distance connectors, but quite defective in the matter of long-distance trunk lines. This means that in case of a traffic jam the processes involving parts of the brain quite remote from one another should suffer first. That is, processes involving several centers, a number of different motor processes, and a considerable number of association areas should be among the least stable in cases of insanity. These are precisely the processes which we should normally class as higher, and we obtain another confirmation of our expectation, which seems to be verified by experience, that the higher processes deteriorate first in insanity.

There is some evidence that the long-distance paths in the brain have a tendency to run outside of the cerebrum altogether and to traverse the lower centers. This is indicated by the remarkably small damage done by cutting some of the long-distance cerebral loops of white matter. It almost seems as if these superficial connections were so inadequate that they furnish only a small part of the connections really needed.

With reference to this, the phenomena of handedness and of hemispheric dominance are interesting. Handedness seems to occur in the lower mammals, though it is less conspicuous than in man, probably in part because of the lower degree of organization and skill demanded by the tasks which they perform. Nevertheless, the choice between the right and the left side in muscular skill does actually seem to be less than in man even in the lower primates.

The right-handedness of the normal man, as is well known, is generally associated with a left-brainedness, and the left-handedness of a minority of humans with a right-brainedness. That is, the cerebral functions are not distributed evenly over the two hemispheres, and one of these, the dominant hemisphere, has the lion's share of the higher functions. It is true that many essentially bilateral functions—those involving the fields of vision, for example— are represented each in its appropriate hemisphere, though this is not true for *all* bilateral functions. However, most of the "higher" areas are confined to the dominant hemisphere. For example, in the adult, the effect of an extensive injury in the secondary hemisphere is far less serious than the effect of a similar injury in the dominant hemisphere. At a relatively early age in his career, Pasteur suffered a cerebral hemorrhage on his right side which left him with a moderate degree of one-sided paralysis, a hemiplegia. When he died, his brain was examined, and he was found to be suffering from a right-sided injury, so extensive that it has been said that after his injury "he had only half a brain." There certainly were extensive lesions of the parietal and temporal regions. Nevertheless, after this injury he did some of his best work. A similar injury of the left side in a right-handed adult would almost certainly have been fatal and would certainly reduce the patient into an animal condition of mental and nervous crippledness.

It is said that the situation is considerably better in early infancy, and that in the first six months of life an extensive injury to the dominant hemisphere may compel the normally secondary hemisphere to take its place; so that the patient appears far more nearly normal than he would be had the injury occurred at a later stage. This is quite in accordance with the general great flexibility shown by the nervous system in the early weeks of life, and the great rigidity which it rapidly develops later. It is possible that, short of such serious injuries, handedness is reasonably flexible in the very young child. However, long before the child is of school age, the natural handedness and cerebral dominance are established for life. It used to be thought that left-handedness was a serious social disadvantage. With most tools, school desks, and sports equipment primarily made for the right-handed, it certainly is to some extent. In the past, moreover, it was viewed with some of the superstitious disapproval that has attached to so many minor variations from the human norm, such as birthmarks or red hair. From a combination of motives, many people have attempted and even succeeded, in changing the external handedness of their children by education, though of course they could not change its physiological basis in hemispheric dominance. It was then found that in very many cases these hemispheric changelings suffered from stuttering and other defects of speech, reading, and writing, to the extent of seriously wounding their prospects in life and their hopes for a normal career.

We now see at least one possible explanation for the phenomenon. With the education of the secondary hand, there has been a partial education of that part of the secondary hemisphere which deals with skilled motions, such as writing. Since, however, these motions are carried out in the closest possible association with reading, speech, and other activities which are inseparably connected with the dominant hemisphere, the neuron chains involved in processes of the sort must cross over from hemisphere to hemisphere and back; and in a process of any complication, they must do this again and again. Now, the direct connectors between the hemispheres—the cerebral commissures—in a brain as large as that of man are so few in number that they are of very little use, and the interhemispheric traffic must go by roundabout routes through the brain stem, which we know very imperfectly but which are certainly long, scanty, and subject to interruption. As a consequence, the processes associated with speech and writing are very likely to be involved in a traffic jam, and stuttering is the most natural thing in the world.

That is, the human brain is probably too large already to use in an efficient manner all the facilities which seem to be anatomically present. In a cat, the destruction of the dominant hemisphere seems to produce relatively less damage than in man, and the destruction of the secondary hemisphere probably more damage. At any rate, the apportionment of function in the two hemispheres is more nearly equal. In man, the gain achieved by the increase in size and complication of the brain is partly nullified by the fact that less of the organ can be used effectively at one time. It is interesting to reflect that we may be facing one of those limitations of nature in which highly specialized organs reach a level of declining efficiency and ultimately lead to the extinction of the species. The human brain may be as far along on its road to this destructive specialization as the great nose horns of the last of the titanotheres.

VIII

Information, Language, and Society

The concept of an organization, the elements of which are themselves small organizations, is neither unfamiliar nor new. The loose federations of ancient Greece, the Holy Roman Empire and its similarly constituted feudal contemporaries, the Swiss Companions of the Oath, the United Netherlands, the United States of America, and the many United States to the south of it, the Union of Socialist Soviet Republics, are all examples of hierarchies of organizations on the political sphere. The Leviathan of Hobbes, the Man-State made up of lesser men, is an illustration of the same idea one stage lower in scale, while Leibniz's treatment of the living organism as being really a plenum, wherein other living organisms, such as the blood corpuscles, have their life, is but another step in the same direction. It is, in fact, scarcely more than a philosophical anticipation of the cell theory, according to which most of the animals and plants of moderate size and all of those of large dimensions are made up of units, cells, which have many if not all the attributes of independent living organism. The multicellular organisms may themselves be the building bricks of organisms of a higher stage, such as the Portuguese man-of-war, which is a complex structure of differentiated coelenterate polyps, where the several individuals are modified in different ways to serve the nutrition, the support, the locomotion, the excretion, the reproduction, and the support of the colony as a whole.

Strictly speaking, such a physically conjoint colony as that poses no question of organization which is philosophically deeper than those which arise at a lower level of individuality. It is very different with man and the other social animals—with the herds of baboons or cattle, the beaver colonies, the hives of bees, the nests of wasps or ants. The degree of integration of the life of the community may very well approach the level shown in the conduct of a single individual, yet the individual will probably have a fixed nervous system, with permanent topographic relations between the elements and permanent connections, while the community consists of individuals with shifting relations in space and time and no permanent, unbreakable physical connections. All the nervous tissue of the beehive is the nervous tissue of some single bee. How then does the beehive act in unison, and at that in a very variable, adapted, organized unison? Obviously, the secret is in the intercommunication of its members.

This intercommunication can vary greatly in complexity and content. With man, it embraces the whole intricacy of language and literature, and very much besides. With the ants, it probably does not cover much more than a few smells. It is very improbable that an ant can distinguish one ant from another. It certainly can distinguish an ant from its own nest from an ant from a foreign nest, and may cooperate with the one, destroy the other. Within a few outside reactions of this kind, the ant seems to have a mind almost as patterned, chitin-bound, as its body. It is what we might expect *a priori* from an animal whose growing phase and, to a large extent, whose learning phase are rigidly separated from the phase of mature activity. The only means of communication we can trace in them are as general and diffuse as the hormonal system of communication within the body. Indeed, smell, one of the chemical senses, general and undirectional as it is, is not unlike the hormonal influences within the body.

Let it be remarked parenthetically that musk, civet, castoreum, and the like sexually attractive substances in the mammals may be regarded as communal, exterior hormones, indispensable, especially in solitary animals, for the bringing the sexes together at the proper time, and serve for the continuation of the race. By this I do not mean to assert that the inner action of these substances, once they reach the organ of smell, is hormonal rather than nervous. It is hard to see how it can be purely hormonal in quantities as small as those which are readily perceivable; on the other hand, we know too little of the action of the hormones to deny the possibility of the hormonal action of vanishingly small quantities of such substances. Moreover, the long, twisted rings of carbon atoms found in muskone and civetone do not need too much rearrangement to form the linked ring structure characteristic of the sex hormones, some of the vitamins, and some of the carcinogens.   I do not care to pronounce an opinion on this matter; I leave it as an interesting speculation.

The odors perceived by the ant seem to lead to a highly standardized course of conduct; but the value of a simple stimulus, such as an odor, for conveying information depends not only on the information conveyed by the stimulus itself but on the whole nervous constitution of the sender and the receiver of the stimulus as well.   Suppose I find myself in the woods with an intelligent savage who cannot speak my language and whose language I cannot speak.   Even without any code of sign language common to the two of us, I can learn a great deal from him.   All I need to do is to be alert to those moments when he shows the signs of emotion or interest.   I then cast my eyes around, perhaps paying special attention to the direction of his glance, and fix in my memory what I see or hear.   It will not be long before I discover the things which seem important to him, not because he has communicated them to me by language, but because I myself have observed them.   In other words, a signal without an intrinsic content may acquire meaning in his mind by what he observes at the time, and may acquire meaning in my mind by what I observe at the time.   The ability that he has to pick out the moments of my special, active attention is in itself a language as varied in possibilities as the range of impressions that the two of us are able to encompass.   Thus social animals may have an active, intelligent, flexible means of communication long before the development of language.

Whatever means of communication the race may have, it is possible to define and to measure the amount of information available to the race and to distinguish it from the amount of information available to the individual.   Certainly no information available to the individual is also available to the race unless it modifies the behavior of one individual to another, nor is even that behavior of racial significance unless it is distinguishable by other individuals from other forms of behavior.   Thus the question as to whether a certain piece of information is racial or of purely private availability depends on whether it results in the individual assuming a form of activity which can be recognized as a distinct form of activity by other members of the race, in the sense that it will in turn affect their activity, and so on.

I have spoken of the race.   This is really too broad a term for the scope of most communal information.   Properly speaking, the community extends only so far as there extends an effectual transmission of information. It is possible to give a sort of measure to this by comparing the number of decisions entering a group from outside with the number of decisions made in the group. We can thus measure the autonomy of the group. A measure of the effective size of a group is given by the size which it must have to have achieved a certain stated degree of autonomy.

A group may have more group information or less group information than its members. A group of non-social animals, temporarily assembled, contains very little group information, even though its members may possess much information as individuals. This is because very little that one member does is noticed by the others and is acted on by them in a way that goes further in the group. On the other hand, the human organism contains vastly more information, in all probability, than does any one of its cells. There is thus no necessary relation in either direction between the amount of racial or tribal or community information and the amount of information available to the individual.

As in the case of the individual, not all the information which is available to the race at one time is accessible without special effort. There is a well-known tendency of libraries to become clogged by their own volume; of the sciences to develop such a degree of specialization that the expert is often illiterate outside his own minute specialty. Dr. Vannevar Bush has suggested the use of mechanical aids for the searching through vast bodies of material. These probably have their uses, but they are limited by the impossibility of classifying a book under an unfamiliar heading unless some particular person has already recognized the relevance of that heading for that particular book. In the case where two subjects have the same techniques and intellectual content but belong to widely separated fields, this still requires some individual with an almost Leibnizian catholicity of interest.

In connection with the effective amount of communal information, one of the most surprising facts about the body politic is its extreme lack of efficient homeostatic processes. There is a belief, current in many countries, which has been elevated to the rank of an official article of faith in the United States, that free competition is itself a homeostatic process: that in a free market the individual selfishness of the bargainers, each seeking to sell as high and buy as low as possible, will result in the end in a stable dynamics of prices, and with redound to the greatest common good. This is associated with the very comforting view that the individual entrepreneur, in seeking to forward his own interest, is in some manner a public benefactor and has thus earned the great rewards with which society has showered him.   Unfortunately, the evidence, such as it is, is against this simple-minded theory.   The market is a game, which has indeed received a simulacrum in the family game of Monopoly.   It is thus strictly subject to the general theory of games, developed by von Neumann and Morgenstern.   This theory is based on the assumption that each player, at every stage, in view of the information then available to him, plays in accordance with a completely intelligent policy, which will in the end assure him of the greatest possible expectation of reward.   It is thus the market game as played between perfectly intelligent, perfectly ruthless operators.   Even in the case of two players, the theory is complicated, although it often leads to the choice of a definite line of play.   In many cases, however, where there are three players, and in the overwhelming majority of cases, when the number of players is large, the result is one of extreme indeterminacy and instability.   The individual players are compelled by their own cupidity to form coalitions; but these coalitions do not generally establish themselves in any single, determinate way, and usually terminate in a welter of betrayal, turncoatism, and deception, which is only too true a picture of the higher business life, or the closely related lives of politics, diplomacy, and war.   In the long run, even the most brilliant and unprincipled huckster must expect ruin; but let the hucksters become tired of this and agree to live in peace with one another, and the great rewards are reserved for the one who watches for an opportune time to break his agreement and betray his companions.   There is no homeostasis whatever.   We are involved in the business cycles of boom and failure, in the successions of dictatorship and revolution, in the wars which everyone loses, which are so real a feature of modern times.

Naturally, von Neumann's picture of the player as a completely intelligent, completely ruthless person is an abstraction and a perversion of the facts.   It is rare to find a large number of thoroughly clever and unprincipled persons playing a game together.   Where the knaves assemble, there will always be fools; and where the fools are present in sufficient numbers, they offer a more profitable object of exploitation for the knaves.   The psychology of the fool has become a subject well worth the serious attention of the knaves.   Instead of looking out for his own ultimate interest, after the fashion of von Neumann's gamesters, the fool operates in a manner which, by and large, is as predictable as the struggles of a rat in a maze.   *This* policy of lies—or rather, of statements irrelevant to the truth—will make him buy a particular brand of cigarettes; *that* policy will, or so the party hopes, induce him to vote for a particular candidate—any candidate—or to join in a political witch hunt. A certain precise mixture of religion, pornography, and pseudo science will sell an illustrated newspaper. A certain blend of wheedling, bribery, and intimidation will induce a young scientist to work on guided missiles or the atomic bomb. To determine these, we have our machinery of radio fan ratings, straw votes, opinion samplings, and other psychological investigations, with the common man as their object; and there are always the statisticians, sociologists, and economists available to sell their services to these undertakings.

Luckily for us, these merchants of lies, these exploiters of gullibility, have not yet arrived at such a pitch of perfection as to have things all their own way. This is because no man is either all fool or all knave. The average man is quite reasonably intelligent concerning subjects which come to his direct attention and quite reasonably altruistic in matters of public benefit or private suffering which are brought before his own eyes. In a small country community which has been running long enough to have developed somewhat uniform levels of intelligence and behavior, there is a very respectable standard of care for the unfortunate, of administration of roads and other public facilities, of tolerance for those who have offended once or twice against society. After all, these people are there, and the rest of the community must continue to live with them. On the other hand, in such a community, it does not do for a man to have the habit of overreaching his neighbors. There are ways of making him feel the weight of public opinion. After a while, he will find it so ubiquitous, so unavoidable, so restricting and oppressing that he will have to leave the community in self-defense.

Thus small, closely knit communities have a very considerable measure of homeostasis; and this, whether they are highly literate communities in a civilized country or villages of primitive savages. Strange and even repugnant as the customs of many barbarians may seem to us, they generally have a very definite homeostatic value, which it is part of the function of anthropologists to interpret. It is only in the large community, where the Lords of Things as They Are protect themselves from hunger by wealth, from public opinion by privacy and anonymity, from private criticism by the laws of libel and the possession of the means of communication, that ruthlessness can reach its most sublime levels. Of all of these anti-homeostatic factors in society, the control of the means of communication is the most effective and most important.

One of the lessons of the present book is that any organism is held together in this action by the possession of means for the acquisition, use, retention, and transmission of information. In a society too large for the direct contact of its members, these means are the press, both as it concerns books and as it concerns newspapers, the radio, the telephone system, the telegraph, the posts, the theater, the movies, the schools, and the church. Besides their intrinsic importance as means of communication, each of these serves other, secondary functions. The newspaper is a vehicle for advertisement and an instrument for the monetary gain of its proprietor, as are also the movies and the radio. The school and the church are not merely refuges for the scholar and the saint: they are also the home of the Great Educator and the Bishop. The book that does not earn money for its publisher probably does not get printed and certainly does not get reprinted.

In a society like ours, avowedly based on buying and selling, in which all natural and human resources are regarded as the absolute property of the first business man enterprising enough to exploit them, these secondary aspects of the means of communication tend to encroach further and further on the primary ones. This is aided by the very elaboration and the consequent expense of the means themselves. The country paper may continue to use its own reporters to canvass the villages around for gossip, but it buys its national news, its syndicated features, its political opinions, as stereotyped "boiler plate." The radio depends on its advertisers for income, and, as everywhere, the man who pays the piper calls the tune. The great news services cost too much to be available to the publisher of moderate means. The book publishers concentrate on books that are likely to be acceptable to some book club which buys out the whole of an enormous edition. The college president and the Bishop, even if they have no personal ambitions for power, have expensive institutions to run and can only seek their money where the money is.

Thus on all sides we have a triple constriction of the means of communication: the elimination of the less profitable means in favor of the more profitable; the fact that these means are in the hands of the very limited class of wealthy men, and thus naturally express the opinions of that class; and the further fact that, as one of the chief avenues to political and personal power, they attract above all those ambitious for such power. That system which more than all others should contribute to social homeostasis is thrown directly into the hands of those most concerned in the game of power and money, which we have already seen to be one of the chief anti-homeostatic elements in the community. It is no wonder then that the larger communities, subject to this disruptive influence, contain far less communally available information than the smaller communities, to say nothing of the human elements of which all communities are built up. Like the wolf pack, although let us hope to a lesser extent, the State is stupider than most of its components.

This runs counter to a tendency much voiced among business executives, heads of great laboratories, and the like, to assume that because the community is larger than the individual it is also more intelligent. Some of this opinion is due to no more than a childish delight in the large and the lavish. Some of it is due to a sense of the possibilities of a large organization for good. Not a little of it, however, is nothing more than an eye for the main chance and a lusting after the fleshpots of Egypt.

There is another group of those who see nothing good in the anarchy of modern society, and in whom an optimistic feeling that there must be some way out has led to an overvaluation of the possible homeostatic elements in the community. Much as we may sympathize with these individuals and appreciate the emotional dilemma in which they find themselves, we cannot attribute too much value to this type of wishful thinking. It is the mode of thought of the mice when faced with the problem of belling the cat. Undoubtedly it would be very pleasant for us mice if the predatory cats of this world were to be belled, but—who is going to do it? Who is to assure us that ruthless power will not find its way back into the hands of those most avid for it?

I mention this matter because of the considerable, and I think false, hopes which some of my friends have built for the social efficacy of whatever new ways of thinking this book may contain. They are certain that our control over our material environment has far outgrown our control over our social environment and our understanding thereof. Therefore, they consider that the main task of the immediate future is to extend to the fields of anthropology, of sociology, of economics, the methods of the natural sciences, in the hope of achieving a like measure of success in the social fields. From believing this necessary, they come to believe it possible. In this, I maintain, they show an excessive optimism, and a misunderstanding of the nature of all scientific achievement.

All the great successes in precise science have been made in fields where there is a certain high degree of isolation of the phenomenon from the observer. We have seen in the case of astronomy that this may result from the enormous scale of certain phenomena with respect to man, so that man's mightiest efforts, not to speak of his mere glance, cannot make the slightest visible impression on the celestial world. In modern atomic physics, on the other hand, the science of the unspeakably minute, it is true that anything we do will have an influence on many individual particles which is great *from the point of view of that particle*. However, we do not live on the scale of the particles concerned, either in space or in time; and the events that might be of the greatest significance from the point of view of an observer conforming to their scale of existence appear to us—with some exceptions, it is true, as in the Wilson cloud-chamber experiments—only as average mass effects in which enormous populations of particles cooperate. As far as these effects are concerned, the intervals of time concerned are large from the point of view of the individual particle and its motion, and our statistical theories have an admirably adequate basis. In short, we are too small to influence the stars in their courses, and too large to care about anything but the mass effects of molecules, atoms, and electrons. In both cases, we achieve a sufficiently loose coupling with the phenomena we are studying to give a massive total account of this coupling, although the coupling may not be loose enough for us to be able to ignore it altogether.

It is in the social sciences that the coupling between the observed phenomenon and the observer is hardest to minimize. On the one hand, the observer is able to exert a considerable influence on the phenomena that come to his attention. With all respect to the intelligence, skill, and honesty of purpose of my anthropologist friends, I cannot think that any community which they have investigated will ever be quite the same afterward. Many a missionary has fixed his own misunderstandings of a primitive language as law eternal in the process of reducing it to writing. There is much in the social habits of a people which is dispersed and distorted by the mere act of making inquiries about it. In another sense from that in which it is usually stated, *traduttore traditore*.

On the other hand, the social scientist has not the advantage of looking down on his subjects from the cold heights of eternity and ubiquity. It may be that there is a mass sociology of the human animalcule, observed like the populations of *Drosophila* in a bottle, but this is not a sociology in which we, who are human animalcules ourselves, are particularly interested. We are not much concerned about human rises and falls, pleasures and agonies, *sub specie aeternitatis.* Your anthropologist reports the customs associated with the life, education, career, and death of people whose life scale is much the same as his own. Your economist is most interested in predicting such business cycles as run their course in less than a generation or, at least, have repercussions which affect a man differentially at different stages of his career. Few philosophers of politics nowadays care to confine their investigations to the world of Ideas of Plato.

In other words, in the social sciences we have to deal with short statistical runs, nor can we be sure that a considerable part of what we observe is not an artifact of our own creation. An investigation of the stock market is likely to upset the stock market. We are too much in tune with the objects of our investigation to be good probes. In short, whether our investigations in the social sciences be statistical or dynamic—and they should participate in the nature of both— they can never be good to more than a very few decimal places, and, in short, can never furnish us with a quantity of verifiable, significant information which begins to compare with that which we have learned to expect in the natural sciences. We cannot afford to neglect them; neither should we build exaggerated expectations of their possibilities. There is much which we must leave, whether we like it or not, to the un-"scientific," narrative method of the professional historian.

<hr />

Note

There is one question which properly belongs to this chapter, though it in no sense represents a culmination of its argument. It is the question whether it is possible to construct a chess-playing machine, and whether this sort of ability represents an essential difference between the potentialities of the machine and the mind. Note that we need not raise the question as to whether it is possible to construct a machine which will play an optimum game in the sense of von Neumann. Not even the best human brain approximates to this. At the other end, it is unquestionably possible to construct a machine that will play chess in the sense of following the rules of the game, irrespective of the merit of the play. This is essentially no more difficult than the construction of a system of interlocking signals for a railway signal tower. The real problem is intermediate: to construct a machine which shall offer interesting opposition to a player at some one of the many levels at which human chess players find themselves.

I think it is possible to construct a relatively crude but not altogether trivial apparatus for this purpose. The machine must actually play—at a high speed if possible—all its own admissible moves and all the opponent's admissible ripostes for two or three moves ahead. To each sequence of moves it should assign a certain conventional valuation. Here, to checkmate the opponent receives the highest valuation at each stage, to be checkmated, the lowest; while losing pieces, taking opponent's pieces, checking, and other recognizable situations should receive valuations not too remote from those which good players would assign them. The first of an entire sequence of moves should receive a valuation much as von Neumann's theory would assign it. At the stage at which the machine is to play once and the opponent once, the valuation of a play by the machine is the minimum valuation of the situation after the opponent has made all possible plays. At the stage where the machine is to play twice and the opponent twice, the valuation of a play by the machine is the minimum with respect to the opponent's first play of the maximum valuation of the plays by the machine at the stage when there is only one play of the opponent and one by the machine to follow. This process can be extended to the case when each player makes three plays, and so on. Then the machine chooses any one of the plays giving the maximum valuation for the stage n plays ahead, where n has some value on which the designer of the machine has decided. This it makes as its definitive play.

Such a machine would not only play legal chess, but a chess not so manifestly bad as to be ridiculous. At any stage, if there were a mate possible in two or three moves, the machine would make it; and if it were possible to avoid a mate by the enemy in two or three moves, the machine would avoid it. It would probably win over a stupid or careless chess player, and would almost certainly lose to a careful player of any considerable degree of proficiency. In other words, it might very well be as good a player as the vast majority of the human race. This does not mean that it would reach the degree of proficiency of Maelzel's fraudulent machine, but, for all that, it may attain a pretty fair level of accomplishment.

PART II

SUPPLEMENTARY CHAPTERS

1 9 6 1

IX

On Learning
and Self-Reproducing Machines

Two of the phenomena which we consider to be characteristic of living systems are the power to learn and the power to reproduce themselves. These properties, different as they appear, are intimately related to one another. An animal that learns is one which is capable of being transformed by its past environment into a different being and is therefore adjustable to its environment within its individual lifetime. An animal that multiplies is able to create other animals in its own likeness at least approximately, although not so completely in its own likeness that they cannot vary in the course of time. If this variation is itself inheritable, we have the raw material on which natural selection can work. If the hereditary invariability concerns manners of behavior, then among the varied patterns of behavior which are propagated some will be found advantageous to the continuing existence of the race and will establish themselves, while others which are detrimental to this continuing existence will be eliminated. The result is a certain sort of racial or phylogenetic learning, as contrasted with the ontogenetic learning of the individual. Both ontogenetic and phylogenetic learning are modes by which the animal can adjust itself to its environment.

Both ontogenetic and phylogenetic learning, and certainly the latter, extend themselves not only to all animals but to plants and, indeed, to all organisms which in any sense may be considered to be living. However, the degree to which these two forms of learning are found to be important in different sorts of living beings varies widely. In man, and to a lesser extent in the other mammals, ontogenetic learning and individual adaptability are raised to the highest point.   Indeed, it may be said that a large part of the phylogenetic learning of man has been devoted to establishing the possibility of good ontogenetic learning.

It has been pointed out by Julian Huxley in his fundamental paper on the mind of birds[1] that birds have a small capacity for ontogenetic learning.   Something similar is true in the case of insects, and in both instances it may be associated with the terrific demands made on the individual by flight and the consequential pre-emption of the capabilities of the nervous system which might otherwise be applied to ontogenetic learning.   Complicated as the behavior patterns of birds are—in flying, in courtship, in the care of the young, and in nest building—they are carried out correctly the very first time without the need of any large amount of instruction from the mother.

It is altogether appropriate to devote a chapter of this book to these two related subjects.   Can man-made machines learn and can they reproduce themselves?   We shall try to show in this chapter that in fact they can learn and can reproduce themselves, and we shall give an account of the technique needed for both these activities.

The simpler of these two processes is that of learning, and it is there that the technical development has gone furthest.   I shall talk here particularly of the learning of game-playing machines which enables them to improve the strategy and tactics of their performance by experience.

There is an established theory of the playing of games—the von Neumann theory.[2]   It concerns a policy which is best considered by working from the end of the game rather than from the beginning. In the last move of the game, a player strives to make a winning move if possible, and if not, then at least a drawing move.   His opponent, at the previous stage, strives to make a move which will prevent the other player from making a winning or a drawing move. If he can himself make a winning move at that stage, he will do so, and this will not be the next to the last but the last stage of the game. The other player at the move before this will try to act in such a way that the very best resources of his opponent will not prevent him from ending with a winning move, and so on backward.

There are games such as ticktacktoe where the entire strategy is known, and it is possible to start this policy from the very be-

[1] Huxley, J., *Evolution: The Modern Synthesis*, Harper Bros., New York, 1943.

[2] von Neumann, J., and O. Morgenstern, *Theory of Games and Economic Behavior*, Princeton University Press, Princeton, N.J., 1944.

ginning. When this is feasible, it is manifestly the best way of playing the game. However, in many games like chess and checkers our knowledge is not sufficient to permit a complete strategy of this sort, and then we can only approximate to it. The von Neumann type of approximate theory tends to lead a player to act with the utmost caution, assuming that his opponent is the perfectly wise sort of a master.

This attitude, however, is not always justified. In war, which is a sort of game, this will in general lead to an indecisive action which will often be not much better than a defeat. Let me give two historical examples. When Napoleon fought the Austrians in Italy, it was part of his effectiveness that he knew the Austrian mode of military thought to be hidebound and traditional, so that he was quite justified in assuming that they were incapable of taking advantage of the new decision-compelling methods of war which had been developed by the soldiers of the French Revolution. When Nelson fought the combined fleets of continental Europe, he had the advantage of fighting with a naval machine which had kept the seas for years and which had developed methods of thought of which, as he was well aware, his enemies were incapable. If he had not made the fullest possible use of this advantage, instead of acting as cautiously as he would have had to act under the supposition that he was facing an enemy of equal naval experience, he might have won in the long run but could not have won so quickly and decisively as to establish the tight naval blockade which was the ultimate downfall of Napoleon. Thus, in both cases, the guiding factor was the known record of the commander and of his opponents, as exhibited statistically in the past of their actions, rather than an attempt to play the perfect game against the perfect opponent. Any direct use of the von Neumann method of game theory in these cases would have proved futile.

In a similar way, books on chess theory are not written from the von Neumann point of view. They are compendia of principles drawn from the practical experience of chess players playing against other chess players of high quality and wide knowledge; and they establish certain values or weightings to be given to the loss of each piece, to mobility, to command, to development, and to other factors which may vary with the stage of the game.

It is not very difficult to make machines which will play chess of a sort. The mere obedience to the laws of the game, so that only legal moves are made, is easily within the power of quite simple computing machines. Indeed, it is not hard to adapt an ordinary digital machine to these purposes.

Now comes the question of policy within the rules of the game. Every evaluation of pieces, command, mobility, and so forth, is intrinsically capable of being reduced to numerical terms; and when this is done, the maxims of a chess book may be used for the determination of the best moves of each stage. Such machines have been made; and they will play a very fair amateur chess, although at present not a game of master caliber.

Imagine yourself in the position of playing chess against such a machine. To make the situation fair, let us suppose you are playing correspondence chess without the knowledge that it is such a machine you are playing and without the prejudices that this knowledge may excite. Naturally, as always is the case with chess, you will come to a judgment of your opponent's chess personality. You will find that when the same situation comes up twice on the chessboard, your opponent's reaction will be the same each time, and you will find that he has a very rigid personality. If any trick of yours will work, then it will always work under the same conditions. It is thus not too hard for an expert to get a line on his machine opponent and to defeat him every time.

However, there are machines that cannot be defeated so trivially. Let us suppose that every few games the machine takes time off and uses its facilities for another purpose. This time, it does not play against an opponent, but examines all the previous games which it has recorded on its memory to determine what weighting of the different evaluations of the worth of pieces, command, mobility, and the like, will conduce most to winning. In this way, it learns not only from its own failures but its opponent's successes. It now replaces its earlier valuations by the new ones and goes on playing as a new and better machine. Such a machine would no longer have as rigid a personality, and the tricks which were once successful against it will ultimately fail. More than that, it may absorb in the course of time something of the policy of its opponents.

All this is very difficult to do in chess, and as a matter of fact the full development of this technique, so as to give rise to a machine that can play master chess, has not been accomplished. Checkers offers an easier problem. The homogeneity of the values of the pieces greatly reduces the number of combinations to be considered. Moreover, partly as a consequence of this homogeneity, the checker game is much less divided into distinct stages than the chess game. Even in checkers, the main problem of the end game is no longer to take pieces but to establish contact with the enemy so that one is in a position to take pieces. Similarly, the valuation of moves in the chess game must be made independently for the different stages. Not only is the end game different from the middle game in the considerations which are paramount, but the openings are much more devoted to getting the pieces into a position of free mobility for attack and defense than is the middle game. The result is that we cannot be even approximately content with a uniform evaluation of the various weighting factors for the game as a whole, but must divide the learning process into a number of separate stages. Only then can we hope to construct a learning machine which can play master chess.

The idea of a first-order programming, which may be linear in certain cases, combined with a second-order programming, which uses a much more extensive segment of the past for the determination of the policy to be carried out in the first-order programming, has been mentioned earlier in this book in connection with the problem of prediction. The predictor uses the immediate past of the flight of the airplane as a tool for the prediction of the future by means of a linear operation; but the determination of the correct linear operation is a statistical problem in which the long past of the flight and the past of many similar flights are used to give the basis of the statistics.

The statistical studies necessary to use a long past for a determination of the policy to be adopted in view of the short past are highly non-linear. As a matter of fact, in the use of the Wiener-Hopf equation for prediction,[1] the determination of the coefficients of this equation is carried out in a non-linear manner. In general, a learning machine operates by non-linear feedback. The checker-playing machine described by Samuel[2] and Watanabe[3] can learn to defeat the man that programmed it in a fairly consistent way on the basis of from 10 to 20 operating hours of programming.

Watanabe's philosophical ideas on the use of programming machines are very exciting. On the one hand, he is treating a method of proving an elementary geometrical theorem which shall conform in an optimal way according to certain criteria of elegance and simplicity, as a learning game to be played not against an individual opponent but against what we may call "Colonel Bogey." A similar

[1] Wiener, N., *Extrapolation, Interpolation, and Smoothing of Stationary Time Series with Engineering Applications*, The Technology Press of M.I.T. and John Wiley & Sons, New York, 1949.

[2] Samuel, A. L., "Some Studies in Machine Learning, Using the Game of Checkers," *IBM Journal of Research and Development*, **3**, 210–229 (1959).

[3] Watanabe, S., "Information Theoretical Analysis of Multivariate Correlation," *IBM Journal of Research and Development*, **4**, 66–82 (1960).

game which Watanabe is studying is played in logical induction, when we wish to build up a theory which is optimal in a similar quasi-aesthetic way, on the basis of an evaluation of economy, directness, and the like, by the determination of the evaluation of a finite number of parameters which are left free. This, it is true, is only a limited logical induction, but it is well worth studying.

Many forms of the activity of struggle, which we do not ordinarily consider as games, have a great deal of light thrown on them by the theory of game-playing machines. One interesting example is the fight between a mongoose and a snake. As Kipling points out in "Rikki-Tikki-Tavi," the mongoose is not immune to the poison of the cobra, although it is to some extent protected by its coat of stiff hairs which makes it difficult for the snake to bite home. As Kipling states, the fight is a dance with death, a struggle of muscular skill and agility. There is no reason to suppose that the individual motions of the mongoose are faster or more accurate than those of the cobra. Yet the mongoose almost invariably kills the cobra and comes out of the contest unscathed. How is it able to do this?

I am here giving an account which appears valid to me, from having seen such a fight, as well as motion pictures of other such fights. I do not guarantee the correctness of my observations as interpretations. The mongoose begins with a feint, which provokes the snake to strike. The mongoose dodges and makes another such feint, so that we have a rhythmical pattern of activity on the part of the two animals. However, this dance is not static but develops progressively. As it goes on, the feints of the mongoose come earlier and earlier in phase with respect to the darts of the cobra, until finally the mongoose attacks when the cobra is extended and not in a position to move rapidly. This time the mongoose's attack is not a feint but a deadly accurate bite through the cobra's brain.

In other words, the snake's pattern of action is confined to single darts, each one for itself, while the pattern of the mongoose's action involves an appreciable, if not very long, segment of the whole past of the fight. To this extent the mongoose acts like a learning machine, and the real deadliness of its attack is dependent on a much more highly organized nervous system.

As a Walt Disney movie of several years ago showed, something very similar happens when one of our western birds, the road runner, attacks a rattlesnake. While the bird fights with beak and claws, and a mongoose with its teeth, the pattern of activity is very similar. A bullfight is a very fine example of the same thing. For it must be remembered that the bullfight is not a sport but a dance with death, to exhibit the beauty and the interlaced coordinating actions of the bull and the man.   Fairness to the bull has no part in it, and we can leave out from our point of view the preliminary goading and weakening of the bull, which have the purpose of bringing the contest to a level where the interaction of the patterns of the two participants is most highly developed.   The skilled bullfighter has a large repertory of possible actions, such as the flaunting of the cape, various dodges and pirouettes, and the like, which are intended to bring the bull into a position in which it has completed its rush and is extended at the precise moment that the bullfighter is ready to plunge the *estoque* into the bull's heart.

What I have said concerning the fight between the mongoose and the cobra, or the toreador and the bull, will also apply to physical contests between man and man.   Consider a duel with the small-sword.   It consists of a sequence of feints, parries, and thrusts, with the intention on the part of each participant to bring his opponent's sword out of line to such an extent that he can thrust home without laying himself open to a double encounter.   Again, in a championship game of tennis, it is not enough to serve or return the ball perfectly as far as each individual stroke is considered; the strategy is rather to force the opponent into a series of returns which put him progressively in a worse position until there is no way in which he can return the ball safely.

These physical contests and the sort of games which we have supposed the game-playing machine to play both have the same element of learning in terms of experience of the opponent's habits as well as one's own.   What is true of games of physical encounter is also true of contests in which the intellectual element is stronger, such as war and the games which simulate war, by which our staff officers win the elements of their military experience.   This is true for classical war both on land and at sea, and is equally true with the new and as yet untried war with atomic weapons.   Some degree of mechanization, parallel to the mechanization of checkers by learning machines, is possible in all these.

There is nothing more dangerous to contemplate than World War III.   It is worth considering whether part of the danger may not be intrinsic in the unguarded use of learning machines.   Again and again I have heard the statement that learning machines cannot subject us to any new dangers, because we can turn them off when we feel like it.   But can we?   To turn a machine off effectively, we must be in possession of information as to whether the danger point has come.   The mere fact that we have made the machine does not guarantee that we shall have the proper information to do this. This is already implicit in the statement that the checker-playing machine can defeat the man who has programmed it, and this after a very limited time of working in.   Moreover, the very speed of operation of modern digital machines stands in the way of our ability to perceive and think through the indications of danger.

The idea of non-human devices of great power and great ability to carry through a policy, and of their dangers, is nothing new.   All that is new is that now we possess effective devices of this kind.   In the past, similar possibilities were postulated for the techniques of magic, which forms the theme for so many legends and folk tales. These tales have thoroughly explored the moral situation of the magician.   I have already discussed some aspects of the legendary ethics of magic in an earlier book entitled *The Human Use of Human Beings*.[1] I here repeat some of the material which I have discussed there, in order to bring it out more precisely in its new context of learning machines.

One of the best-known tales of magic is Goethe's "The Sorcerer's Apprentice."   In this, the sorcerer leaves his apprentice and factotum alone with the chore of fetching water.   As the boy is lazy and ingenious, he passes the work over to a broom, to which he has uttered the words of magic which he has heard from his master.   The broom obligingly does the work for him and will not stop.   The boy is on the verge of being drowned out.   He finds that he has not learned, or has forgotten, the second incantation which is to stop the broom.   In desperation, he takes the broomstick, breaks it over his knee, and finds to his consternation that each half of the broom continues to fetch water.   Luckily, before he is completely destroyed, the master returns, says the Words of Power to stop the broom, and administers a good scolding to the apprentice.

Another story is the Arabian Nights tale of the fisherman and the genie.   The fisherman has dredged up in his net a jug closed with the seal of Solomon.   It is one of the vessels in which Solomon has imprisoned the rebellious genie.   The genie emerges in a cloud of smoke, and the gigantic figure tells the fisherman that, whereas in his first years of imprisonment he had resolved to reward his rescuer with power and fortune, he has now decided to slay him out of hand. Luckily for himself, the fisherman finds a way to talk the genie back into the bottle, upon which he casts the jar to the bottom of the ocean.

---

[1] Wiener, N., *The Human Use of Human Beings; Cybernetics and Society*, Houghton Mifflin Company, Boston, 1950.

More terrible than either of these two tales is the fable of the monkey's paw, written by W. W. Jacobs, an English writer of the beginning of the century. A retired English workingman is sitting at his table with his wife and a friend, a returned British sergeant-major from India. The sergeant-major shows his hosts an amulet in the form of a dried, wizened monkey's paw. This has been endowed by an Indian holy man, who has wished to show the folly of defying fate, with the power of granting three wishes to each of three people. The soldier says that he knows nothing of the first two wishes of the first owner, but the last one was for death. He himself, as he tells his friends, was the second owner but will not talk of the horror of his own experiences. He casts the paw into the fire, but his friend retrieves it and wishes to test its powers. His first is for £200. Shortly thereafter there is a knock at the door, and an official of the company by which his son is employed enters the room. The father learns that his son has been killed in the machinery, but that the company, without recognizing any responsibility or legal obligation, wishes to pay the father the sum of £200 as a solatium. The grief-stricken father makes his second wish—that his son may return—and when there is another knock at the door and it is opened, something appears which, we are not told in so many words, is the ghost of the son. The final wish is that this ghost should go away.

In all these stories the point is that the agencies of magic are literal-minded; and that if we ask for a boon from them, we must ask for what we really want and not for what we think we want. The new and real agencies of the learning machine are also literal-minded. If we program a machine for winning a war, we must think well what we mean by winning. A learning machine must be programmed by experience. The only experience of a nuclear war which is not immediately catastrophic is the experience of a war game. If we are to use this experience as a guide for our procedure in a real emergency, the values of winning which we have employed in the programming games must be the same values which we hold at heart in the actual outcome of a war. We can fail in this only at our immediate, utter, and irretrievable peril. We cannot expect the machine to follow us in those prejudices and emotional compromises by which we enable ourselves to call destruction by the name of victory. If we ask for victory and do not know what we mean by it, we shall find the ghost knocking at our door.

So much for learning machines. Now let me say a word or two about self-propagating machines. Here both the words *machine* and *self-propagating* are important. The machine is not only a form of matter, but an agency for accomplishing certain definite purposes. And self-propagation is not merely the creation of a tangible replica; it is the creation of a replica capable of the same functions.

Here, two different points of view come into evidence. One of these is purely combinatorial and concerns the question whether a machine can have enough parts and sufficiently complicated structure to enable self-reproduction to be among its functions. This question has been answered in the affirmative by the late John von Neumann. The other question concerns an actual operative procedure for building self-reproducing machines. Here I shall confine my attentions to a class of machines which, while it does not embrace all machines, is of great generality. I refer to the non-linear transducer.

Such machines are apparatuses which have as an input a single function of time and which have as their output another function of time. The output is completely determined by the past of the input; but in general, the adding of inputs does not add the corresponding outputs. Such pieces of apparatus are known as transducers. One property of all transducers, linear or non-linear, is an invariance with respect to a translation in time. If a machine performs a certain function, then, if the input is shifted back in time, the output is shifted back by the same amount.

Basic to our theory of self-reproducing machines is a canonical form of the representation of non-linear transducers. Here the notions of impedance and admittance, which are so essential in the theory of linear apparatus, are not fully appropriate. We shall have to refer to certain newer methods of carrying out this representation, methods developed partly by me[1] and partly by Professor Dennis Gabor[2] of the University of London.

While both Professor Gabor's methods and my own lead to the construction of non-linear transducers, they are linear to the extent that the non-linear transducer is represented with an output which is the sum of the outputs of a set of non-linear transducers with the same input. These outputs are combined with varying linear coefficients. This allows us to employ the theory of linear developments in the design and specification of the non-linear transducer. And in particular, this method allows us to obtain coefficients of the constituent elements by a least-square process. If we join this to a

[1] Wiener, N., *Nonlinear Problems in Random Theory*, The Technology Press of M.I.T. and John Wiley & Sons, Inc., New York, 1958.

[2] Gabor, D., "Electronic Inventions and Their Impact on Civilization," *Inaugural Lecture*, March 3, 1959, Imperial College of Science and Technology, University of London, England.

method of statistically averaging over the set of all inputs to our apparatus, we have essentially a branch of the theory of orthogonal development. Such a statistical basis of the theory of non-linear transducers can be obtained from an actual study of the past statistics of the inputs used in each particular case.

This is a rough account of Professor Gabor's methods. While mine are essentially similar, the statistical basis for my work is slightly different.

It is well known that electrical currents are not conducted continuously but by a stream of electrons which must have statistical variations from uniformity. These statistical fluctuations can be represented fairly by the theory of the Brownian motion, or by the similar theory of shot effect or tube noise, about which I am going to say something in the next chapter. At any rate, apparatus can be made to generate a standardized shot effect with highly specific statistical distribution, and such apparatus is being manufactured commercially. Note that tube noise is in a sense a universal input in that its fluctuations over a sufficiently long time will sooner or later approximate to any given curve. This tube noise possesses a very simple theory of integration and averaging.

In terms of the statistics of tube noise, we can easily determine a closed set of normal and orthogonal non-linear operations. If the inputs subject to these operations have the statistical distribution appropriate to tube noise, the average product of the output of two component pieces of our apparatus, where this average is taken with respect to the statistical distribution of tube noise, will be zero. Moreover, the mean square output of each apparatus can be normalized to one. The result is that the development of the general non-linear transducer in terms of these components results from an application of the familiar theory of orthonormal functions.

To be specific, our individual pieces of apparatus give outputs which are products of Hermite polynomials in the Laguerre coefficients of the past of the input. This is presented in detail in my *Nonlinear Problems in Random Theory*.

It is of course difficult to average in the first instance over a set of possible inputs. What makes this difficult task realizable is that the shot-effect inputs possess the property known as metric transitivity, or the ergodic property. Any integrable function of the parameter of distribution of shot-effect inputs has in almost every instance a time average equal to its average over the ensemble. This permits us to take two pieces of apparatus with a common shot-effect input, and to determine the average of their product over the entire ensemble of the possible inputs, by taking their product and averaging it over the time. The repertory of operations needed for all these processes involves nothing more than the addition of potentials, the multiplication of potentials, and the operation of averaging over time. Devices exist for all these. As a matter of fact, the elementary devices needed in Professor Gabor's methodology are the same as those needed in mine. One of his students has invented a particularly effective and inexpensive multiplying device depending on the piezo-electric effect on a crystal of the attraction of two magnetic coils.

What this amounts to is that we can imitate any unknown non-linear transducer by a sum of linear terms, each of fixed characteristics and with an adjustable coefficient. This coefficient can be determined as the average product of the outputs of the unknown transducer and a particular known transducer, when the same shot-effect generator is connected to the input of both. What is more, instead of computing this result on the scale of an instrument and then transferring it by hand to the appropriate transducer, thus producing a piecemeal simulation of the apparatus, there is no particular problem in automatically effecting the transfer of the coefficients to the pieces of feedback apparatus. What we have succeeded in doing is to make a white box which can potentially assume the characteristics of any non-linear transducer whatever, and then to draw it into the similitude of a given black-box transducer by subjecting the two to the same random input and connecting the outputs of the structures in the proper manner, so as to arrive at the suitable combination without any intervention on our part.

I ask if this is philosophically very different from what is done when a gene acts as a template to form other molecules of the same gene from an indeterminate mixture of amino and nucleic acids, or when a virus guides into its own form other molecules of the same virus out of the tissues and juices of its host. I do not in the least claim that the details of these processes are the same, but I do claim that they are philosophically very similar phenomena.

X

Brain Waves
and Self-Organizing Systems

In the previous chapter, I discussed the problems of learning and self-propagation as they apply both to machines and, at least by analogy, to living systems. Here I shall repeat certain comments I made in the Preface and which I intend to put to immediate use. As I have pointed out, these two phenomena are closely related to each other, for the first is the basis for the adaptation of the individual to its environment by means of experience, which is what we may call ontogenetic learning, while the second, as it furnishes the material on which variation and natural selection may operate, is the basis of phylogenetic learning. As I have already mentioned, the mammals, in particular man, do a large part of their adjustment to their environment by ontogenetic learning, whereas the birds, with their highly varied patterns of behavior which are not learned in the life of the individual, have devoted themselves much more to phylogenetic learning.

We have seen the importance of non-linear feedbacks in the origination of both processes. The present chapter is devoted to the study of a specific self-organizing system in which non-linear phenomena play a large part. What I here describe is what I believe to be happening in the self-organization of electroencephalograms or brain waves.

Before we can discuss this matter intelligently, I must say something of what brain waves are and how their structure may be subjected to precise mathematical treatment. It has been known for many years that activity of the nervous system is accompanied by

certain electrical potentials.   The first observations in this field go
back to the beginning of the last century and were made by Volta and
Galvani in neuromuscular preparations from the leg of the frog.
This was the birth of the science of electrophysiology.   This science,
however, advanced rather slowly until the end of the first quarter of
the present century.

It is well worth reflecting why the development of this branch of
physiology was so slow.   The original apparatus used for the study
of physiological electric potential consisted of galvanometers.   These
had two weaknesses.   The first was that the entire energy involved
in moving the coil or needle of the galvanometer came from the nerve
itself and was excessively minute.   The second difficulty was that
the galvanometer of those times was an instrument whose mobile
parts had quite appreciable inertia, and a very definite restoring
force was necessary to bring the needle to a well-defined position; that
is, in the nature of the case, the galvanometer was not only a
recording instrument but a distorting instrument.   The best of the
early physiological galvanometers was the string galvanometer of
Einthoven, where the moving parts were reduced to a single wire.
Excellent as this instrument was by the standards of its own time, it
was not good enough to record small electrical potentials without
heavy distortions.

Thus electrophysiology had to wait for a new technique.   This
technique was that of electronics, and took two forms.   One of these
was based on Edison's discovery of certain phenomena pertaining to
the conduction of gases, and from these arose the use of the vacuum
tube or electric valve for amplification.   This made it possible to give
a reasonably faithful transformation of weak potentials into strong
potentials.   And so it permitted us to move the final elements of the
recording device by the use of energy not emanating from the nerve
but controlled by it.

The second invention also involved the conduction of electricity *in
vacuo*, and is known as the cathode-ray oscillograph.   This made it
possible to use as the moving part of the instrument a much lighter
armature than that of any previous galvanometer, namely, a stream
of electrons.   With the aid of these two devices, separately or together, the physiologists of this century have been able to follow
faithfully the time course of small potentials which would have been
completely beyond the range of accurate instrumentation possible in
the nineteenth century.

With these means, we have been able to obtain accurate records of
the time course of the minute potentials arising between two elec-

trodes placed on the scalp or implanted in the brain. While these potentials had already been observed in the nineteenth century, the availability of the new accurate records excited great hopes among the physiologists of twenty or thirty years ago. As to the possibilities of using the devices for the direct study of brain activity, leaders in this field were Berger in Germany, Adrian and Matthews in England, and Jasper, Davis, and the Gibbs (husband and wife) in the United States.

It must be admitted that the later development of electroencephalography has up to now been unable to fulfill the rosy hopes entertained by the early workers in the field. The data which they obtained were recorded by an ink-writer. They are very complicated and irregular curves; and although it was possible to discern certain predominating frequencies, such as the alpha rhythm of about 10 oscillations per second, the ink record was not in a suitable form for further mathematical manipulation. The result is that electroencephalography became more an art than a science, and depended on the ability of the trained observer to recognize certain properties of the ink record on the basis of a large experience. This had the very fundamental objection of making the interpretation of the electroencephalograms a largely subjective matter.

In the late twenties and the early thirties, I had become interested in the harmonic analysis of continuing processes. While the physicists had previously considered such processes, the mathematics of harmonic analysis had been almost confined to the study of either periodic processes or those which in some sense tended to zero as the time became large, positively or negatively. My work was the earliest attempt to put the harmonic analysis of continuing processes on a firm mathematical basis. In this, I found that the fundamental notion was that of autocorrelation, which had already been used by G. I. Taylor (now Sir Geoffrey Taylor) in the study of turbulences.[1]

This autocorrelation for a time function $f(t)$ is represented by the time-mean of the product $f(t + \tau)$ by $f(t)$. It is advantageous to introduce complex functions of the time even though in the actual cases studied we are dealing with real functions. And now the autocorrelation becomes the mean of the product of $f(t + \tau)$ with the conjugate of $f(t)$. Whether we are working with real or complex functions, the power spectrum of $f(t)$ is given by the Fourier transform of the autocorrelation.

I have already spoken of the lack of suitability of ink records for

[1] Taylor, G. I., "Diffusion by Continuous Movements," *Proceedings of the London Mathematical Society*, Ser. 2, **20**, 196–212 (1921–1922).

further mathematical manipulations. Before much could come of the idea of autocorrelation, it was necessary to replace these ink records by other records better adapted to instrumentation.

One of the best ways of recording small fluctuating electric potentials for further manipulation is the use of magnetic tape. This allows the storage of the fluctuating electric potential in a permanent form which can be used later whenever convenient. One such instrument was devised about a decade ago in the Research Laboratory of Electronics of the Massachusetts Institute of Technology, under the guidance of Professor Walter A. Rosenblith and Dr. Mary A. B. Brazier.[1]

In this apparatus, magnetic tape is used in its frequency-modulation form. The reason for this is that the reading of magnetic tape always involves a certain amount of erasure. With amplitude-modulation tape, this erasure gives rise to a change in the message carried, so that in successive readings of the tape we are actually following a changing message.

In frequency modulation there is also a certain amount of erasure, but the instruments by which we read the tape are relatively insensitive to amplitude, and read frequency only. Until the tape is so badly erased that it is completely unreadable, the partial erasure of the tape does not distort appreciably the message which it carries. The result is that the tape can be read very many times with substantially the same accuracy with which it was first read.

As will be seen from the nature of the autocorrelation, one of the tools which we need is a mechanism which will delay the reading of tape by an adjustable amount. If a length of the magnetic tape record having time-duration A is played on an apparatus having two playback heads, one following the other, two signals are generated which are the same except for a relative displacement in time. The time displacement depends on the distance between the playback heads and on the tape speed, and can be varied at will. We can call one of these $f(t)$ and the other $f(t + \tau)$, where τ is the time displacement. The product of the two can be formed, for example, by using square-law rectifiers and linear mixers, and taking advantage of the identity

$$4ab = (a + b)^2 - (a - b)^2 \qquad (10.01)$$

The product can be averaged approximately by integrating with a resistor-capacitor network having a time constant long compared with

[1] Barlow, J. S., and R. M. Brown, *An Analog Correlator System for Brain Potentials*, Technical Report 300, Research Laboratory of Electronics, M.I.T., Cambridge, Mass. (1955).

the duration A of the sample.   The resulting average is proportional
to the value of the autocorrelation function for delay τ.   Repetition
of the process for various values of τ yields a set of values of the
autocorrelation (or rather, the sampled autocorrelation over a large
time base A).   The accompanying graph, Fig. 9, shows a plot of an
actual autocorrelation of this sort.[1]   Let us note that we have shown
only half the curve, for the autocorrelation for negative times would

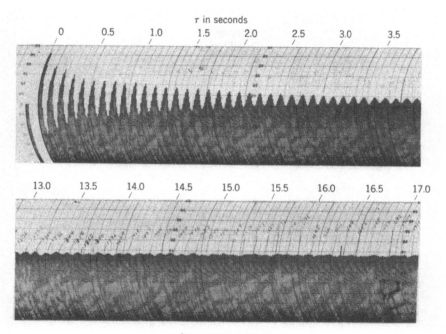

AUTOCORRELATION

FIG. 9.

be the same as that for positive times, at least if the curve of which we
are taking the autocorrelation is real.

Note that similar autocorrelation curves have been used for many
years in optics, and that the instrument by which they have been
obtained is the Michelson interferometer, Fig. 10.   By a system of
mirrors and lenses, the Michelson interferometer divides a beam of
light into two parts which are sent on paths of different length and

[1] This work was undertaken with the cooperation of the Neurophysiology Laboratory of the Massachusetts General Hospital and the Communications Biophysics
Laboratory of M.I.T.

then reunited into one beam. Different path lengths result in different time delays, and the resultant beam is the sum of two replicas of the incoming beam, which may once more be termed $f(t)$ and $f(t + \tau)$. When the beam intensity is measured with a power-sensitive photometer, the reading of the photometer is proportional to the square of $f(t) + f(t + \tau)$, and hence contains a term proportional to the autocorrelation. In other words, the intensity of the interferometer fringes (except for a linear transformation) will give us the autocorrelation.

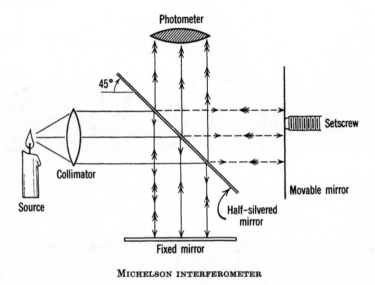

MICHELSON INTERFEROMETER

Fig. 10.

All of this was implicit in Michelson's work. It will be seen that, by carrying out a Fourier transformation on the fringes, the interferometer yields us the power spectrum of the light, and is in fact a spectrometer. It is indeed the most accurate type of spectrometer known to us.

This type of spectrometer has only come into its own in recent years. I am told that it is now accepted as an important tool for precision measurements. The significance of this is that the techniques which I shall now present for the working up of autocorrelation records are equally applicable in spectroscopy and offer methods of pushing to the limit the information which can be yielded by a spectrometer.

Let us discuss the technique of obtaining the spectrum of a brain wave from an autocorrelation. Let $C(t)$ be an autocorrelation of $f(t)$. Then $C(t)$ can be written in the form

$$C(t) = \int_{-\infty}^{\infty} e^{2\pi i \omega t} \, dF(\omega) \qquad (10.02)$$

Here F is always an increasing or at least a non-decreasing function of ω, and we shall term it the integrated spectrum of f. In general, this integrated spectrum is made in three parts, combined additively. The line part of the spectrum increases only at a denumerable set of points.   Take this away, and we are left with a continuous spectrum. This continuous spectrum itself is the sum of two parts, one of which increases only over a set of measure zero, while the other part is absolutely continuous and is the integral of a positive integrable function.

From now on let us suppose that the first two parts of the spectrum —the discrete part and the continuous part which increases over a set of measure zero—are missing.   In this case, we can write

$$C(t) = \int_{-\infty}^{\infty} e^{2\pi i \omega t} \phi(\omega) \, d\omega \qquad (10.03)$$

where $\phi(\omega)$ is the spectral density.   If $\phi(\omega)$ is of Lebesgue class L^2, we can write

$$\phi(\omega) = \int_{-\infty}^{\infty} C(t) e^{-2\pi i \omega t} \, dt \qquad (10.04)$$

As will be seen by looking at the autocorrelation of the brain waves, the predominating part of the power of the spectrum is in the neighborhood of 10 cycles.   In such a case, $\phi(\omega)$ will have a shape similar to the following diagram.

The two peaks near 10 and -10 are mirror images of each other.

The ways of performing a Fourier analysis numerically are various, including the use of integrating instruments and numerical computing processes.   In both cases, it is an inconvenience to the work that the principal peaks are near 10 and -10 and not near 0.   However, there are modes of transferring the harmonic analysis to the neighborhood of zero frequency which greatly cut down the work to be performed.   Notice that

$$\phi(\omega - 10) = \int_{-\infty}^{\infty} C(t)e^{20\pi it}e^{-2\pi i\omega t}\,dt \qquad (10.05)$$

In other words, if we multiply $C(t)$ by $e^{20\pi it}$, our new harmonic analysis will give us a band in the neighborhood of zero frequency and another band in the neighborhood of frequency $+20$.   If we then perform such a multiplication and remove the $+20$ band by averaging methods equivalent to the use of a wave filter, we shall have reduced our harmonic analysis to one in the neighborhood of zero frequency.

Now

$$e^{20\pi it} = \cos 20\pi t + i \sin 20\pi t \qquad (10.06)$$

Therefore, the real and imaginary parts of the $C(t)\cdot e^{20\pi it}$ are given, respectively, by $C(t)\cos 20\pi t$ and $iC(t)\sin 20\pi t$.   The removal of the frequencies in the neighborhood of $+20$ can be performed by putting these two functions through a low-pass filter, which is equivalent to averaging them over an interval of a twentieth of a second or greater.

Suppose that we have a curve where most of the power is nearly at a frequency of 10 cycles.   When we multiply this by the cosine or sine of $20\pi t$, we shall get a curve which is the sum of two parts, one of them behaving locally like this:

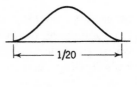

and the other like this:

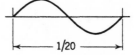

When we average the second curve over the time for a length of a tenth of a second, we get zero.   When we average the first one, we get half of the maximum height.   The result is that, by the smoothing of $C(t)\cos 20\pi t$ and $iC(t)\sin 20\pi t$, we get, respectively, good approximations to the real and imaginary part of a function having all of its frequencies in the neighborhood of zero, and this function will have the distributional frequency around zero that one part of the spectrum of $C(t)$ has around 10. Now let $K_1(t)$ be the result of smoothing $C(t) \cos 20\pi t$ and $K_2(t)$ the result of smoothing $C(t) \sin 20\pi t$. We wish to obtain

$$\int_{-\infty}^{\infty} [K_1(t) + iK_2(t)]e^{-2\pi i\omega t}\, dt$$

$$= \int_{-\infty}^{\infty} [K_1(t) + iK_2(t)][\cos 2\pi\omega t - i\sin 2\pi\omega t]\, dt \qquad (10.07)$$

This expression must be real, since it is a spectrum. Therefore, it will equal

$$\int_{-\infty}^{\infty} K_1(t) \cos 2\pi\omega t\, dt + \int_{-\infty}^{\infty} K_2(t) \sin 2\pi\omega t\, dt \qquad (10.08)$$

In other words, if we make a cosine analysis of K_1 and a sine analysis of K_2, and add them together, we shall have the displaced spectrum of f. It can be shown that K_1 will be even and K_2 will be odd. This means that if we do a cosine analysis of K_1 and add or subtract the sine analysis of K_2, we shall obtain, respectively, the spectrum to the right and to the left of the central frequency at the distance ω. This method for obtaining the spectrum we shall describe as the method of heterodyning.

In the case of autocorrelations which are locally nearly sinusoidal of period, say 0.1 (such as that which appears in the brain-wave autocorrelation of Fig. 9), the computation involved in this method of heterodyning may be simplified. We take our autocorrelation at intervals of a fortieth of a second. We then take the sequence at 0, 1/20 second, 2/20 second, 3/20 second, and so on, and change the sign of those fractions with odd numerators. We average these consecutively for a suitable length of run and get a quantity nearly equal to $K_1(t)$. If we work similarly with the values at 1/40 second, 3/40 second, 5/40 second, and so on, changing the sign of alternate quantities, and perform the same averaging process as before, we get an approximation to $K_2(t)$. From this stage on the procedure is clear.

The justification for this procedure is that the distribution of mass which is

  1 at points $2\pi n$
  -1 at points $(2n + 1)\pi$

while it is zero elsewhere, when it is subject to a harmonic analysis, will contain a cosine component of frequency 1 and no sine component. Similarly, a distribution of mass which is

$$1 \text{ at } (2n + 1/2)\pi$$
$$-1 \text{ at } (2n - 1/2)\pi$$

and

      0 elsewhere

will contain the sine component of frequency 1 and no cosine component. Both distributions will also contain components of frequencies N; but since the original curve which we are analyzing is wanting or nearly wanting at these frequencies, these terms will produce no effect. This greatly simplifies our heterodyning, because the only factors which we have to multiply by are $+1$ or -1.

We have found this method of heterodyning very useful in the harmonic analysis of brain waves when we have only manual means at our disposal, and when the bulk of the work becomes overwhelming if we carry through all the details of the harmonic analysis without the use of heterodyning. All of our earlier work with the harmonic analysis of brain spectra has been done by the heterodyning method. Since, however, it later proved possible to obtain the use of a digital computer for which reducing the bulk of the work is not such a serious consideration, much of our later work in harmonic analysis has been done directly without the use of heterodyning. There will still be much work to be done in places where digital computers are not available, so that I do not consider the heterodyning method obsolete in practice.

I am presenting here portions of a specific autocorrelation which we have obtained in our work. Since the autocorrelation covers a great length of data, it is not suitable for reproducing as a whole here, and we give merely the beginning, in the neighborhood $\tau = 0$, and a portion of it further out.

Figure 11 represents the results of a harmonic analysis of the autocorrelation of which part is exhibited in Fig. 9. In this case, our result was obtained with a high-speed digital computer,[1] but we have found a very good concordance between this spectrum and the one we obtained earlier through heterodyning methods by hand, at least in the neighborhood of the strong part of the spectrum.

When we inspect the curve, we find a remarkable drop in power in the neighborhood of frequency 9.05 cycles per second. The point at which the spectrum substantially fades out is very sharp and gives

---

[1] The IBM-709 at the M.I.T. Computation Center was used.

an objective quantity which can be verified with much greater accuracy than any quantity so far occurring in electroencephalography. There is a certain amount of indication that in other curves which we have obtained, but which are of somewhat questionable reliability in their details, this sudden fall-off in power is followed quite shortly by a sudden rise, so that between them we have a dip in the curve. Whether this be the case or not, there is a strong suggestion that the power in the peak corresponds to a pulling of the power away from the region where the curve is low.

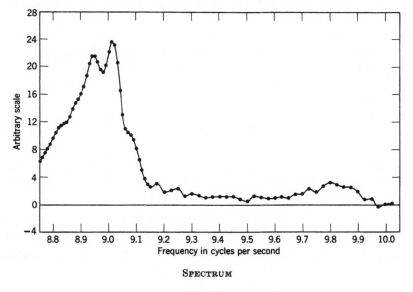

SPECTRUM

FIG. 11.

In the spectrum which we have obtained, it is worth noting that the overwhelming part of the peak lies within a range of about a third of a cycle. An interesting thing is that with another electro-encephalogram of the same subject recorded four days later, this approximate width of the peak is retained and there is more than a suggestion that the form is retained in some detail. There is also reason to believe that with other subjects the width of the peak will be different and perhaps narrower. A thoroughly satisfactory verification of this waits for investigations yet to be made.

It is highly desirable that the sort of work which we have mentioned in these suggestions be followed up by more accurate instrumental work with better instruments so that the suggestions which we here make can be definitely verified or definitely rejected.

I now wish to take up the sampling problem. For this I shall have to introduce some ideas from my previous work on integration in function space.[1] With the aid of this tool, we shall be able to construct a statistical model of a continuing process with a given spectrum. While this model is not an exact replica of the process that generates brain waves, it is near enough to it to yield statistically significant information concerning the root-mean-square error to be expected in brain-wave spectra such as the one already presented in this chapter.

I here state without proof some properties of a certain real function $x(t, \alpha)$ already stated in my paper on generalized harmonic analysis and elsewhere.[1] The real function $x(t, \alpha)$ is dependent on a variable t running from $-\infty$ to ∞ and a variable α running from 0 to 1. It represents one space variable of a Brownian motion dependent on the time t and the parameter α of a statistical distribution. The expression

$$\int_{-\infty}^{\infty} \phi(t) \, dx(t, \alpha) \tag{10.09}$$

is defined for all functions $\phi(t)$ of Lebesgue class L^2 from $-\infty$ to ∞. If $\phi(t)$ has a derivative belonging to L^2, Expression 10.09 is defined as

$$- \int_{-\infty}^{\infty} x(t, \alpha) \phi'(t) \, dt \tag{10.10}$$

and is then defined for all functions $\phi(t)$ belonging to L^2 by a certain well-defined limit process. Other integrals

$$\int_{-\infty}^{\infty} \cdots \int_{-\infty}^{\infty} K(\tau_1, \cdots \tau_n) \, dx(\tau_1, \alpha) \cdots dx(\tau_n, \alpha) \tag{10.11}$$

are defined in a similar manner. The fundamental theorem of which we make use is that

$$\int_0^1 d\alpha \int_{-\infty}^{\infty} \cdots \int_{-\infty}^{\infty} K(\tau_1, \cdots, \tau_n) \, dx(\tau_1, \alpha) \cdots dx(\tau_n, \alpha) \tag{10.12}$$

is obtained by putting

$$K_1(\tau_1, \cdots, \tau_{n/2}) = \sum K(\sigma_1, \sigma_2, \cdots, \sigma_n) \tag{10.13}$$

---

[1] Wiener, N., "Generalized Harmonic Analysis," *Acta Mathematica*, **55**, 117–258 (1930); *Nonlinear Problems in Random Theory*, The Technology Press of M.I.T. and John Wiley & Sons, Inc., New York, 1958.

where the τ_k are formed in all possible ways by identifying all pairs of the σ_k with each other (if n is even), and forming

$$\int_{-\infty}^{\infty} \cdots \int_{-\infty}^{\infty} K_1\Big(\tau_1, \cdots, \tau_{n/2}\Big) d\tau_1, \cdots, d\tau_{n/2} \qquad (10.14)$$

If n is odd,

$$\int_0^1 d\alpha \int_{-\infty}^{\infty} \cdots \int_{-\infty}^{\infty} K(\tau_1, \cdots, \tau_n)\, dx(\tau_1, \alpha) \cdots dx(\tau_n, \alpha) = 0 \quad (10.15)$$

Another important theorem concerning these stochastic integrals is that if $\mathscr{F}\{g\}$ is a functional of $g(t)$, such that $\mathscr{F}[x(t, \alpha)]$ is a function belonging to L in α and depending only on the differences $x(t_2, \alpha) - x(t_1, \alpha)$, then for each t_1 for almost all values of α

$$\lim_{A \to \infty} \frac{1}{A} \int_0^A \mathscr{F}[x(t, \alpha)]dt = \int_0^1 \mathscr{F}[x(t_1, \alpha)]\, d\alpha \qquad (10.16)$$

This is the ergodic theorem of Birkhoff, and has been proved by the author[1] and others.

It has been established in the *Acta Mathematica* paper already mentioned that if U is a real unitary transformation of the function $K(t)$,

$$\int_{-\infty}^{\infty} UK(t)\, dx(t, \alpha) = \int_{-\infty}^{\infty} K(t)\, dx(t, \beta) \qquad (10.17)$$

where β differs from α only by a measure-preserving transformation of the interval $(0, 1)$ into itself.

Now let $K(t)$ belong to L², and let

$$K(t) = \int_{-\infty}^{\infty} q(\omega)e^{2\pi i \omega t}\, d\omega \qquad (10.18)$$

in the Plancherel[2] sense.   Let us examine the real function

$$f(t, \alpha) = \int_{-\infty}^{\infty} K(t + \tau)\, dx(\tau, \alpha) \qquad (10.19)$$

which represents the response of a linear transducer to a Brownian

---

[1] Wiener, N., "The Ergodic Theorem," *Duke Mathematical Journal*, **5**, 1–39 (1939); also in *Modern Mathematics for the Engineer*, E. F. Beckenbach (Ed.), McGraw-Hill, New York, 1956, pp. 166–168.

[2] Wiener, N., "Plancherel's Theorem," *The Fourier Integral and Certain of Its Applications*, The University Press, Cambridge, England, 1933, pp. 46–71; Dover Publications, Inc., New York.

input. This will have the autocorrelation

$$\lim_{T\to\infty} \frac{1}{2T} \int_{-T}^{T} f(t + \tau, \alpha)\overline{f(t, \alpha)}\, dt \qquad (10.20)$$

and this, by the ergodic theorem, will have for almost all values of α the value

$$\int_{0}^{1} d\alpha \int_{-\infty}^{\infty} K(t_1 + \tau)\, dx(t_1, \alpha) \int_{-\infty}^{\infty} \overline{K(t_2)}\, dx(t_2, \alpha)$$

$$= \int_{-\infty}^{\infty} K(t + \tau)\overline{K(t)}\, dt \quad (10.21)$$

The spectrum will then almost always be

$$\int_{-\infty}^{\infty} e^{-2\pi i \omega \tau}\, d\tau \int_{-\infty}^{\infty} K(t + \tau)\overline{K(t)}\, dt$$

$$= \left| \int_{-\infty}^{\infty} K(\tau)e^{-2\pi i \omega \tau}\, d\tau \right|^2$$

$$= |q(\omega)|^2 \qquad (10.22)$$

This is, however, the true spectrum. The sampled autocorrelation over the averaging time A (in our case 2700 seconds) will be

$$\frac{1}{A} \int_{0}^{A} f(t + \tau, \alpha)\overline{f(t, \alpha)}\, dt$$

$$= \int_{-\infty}^{\infty} dx(t_1, \alpha) \int_{-\infty}^{\infty} dx(t_2, \alpha) \frac{1}{A} \int_{0}^{A} K(t_1 + \tau + s)\overline{K(t_2 + s)}\, ds$$

$$(10.23)$$

The resulting sampled spectrum will almost always have the time average

$$\int_{-\infty}^{\infty} e^{-2\pi i \omega \tau}\, d\tau \frac{1}{A} \int_{0}^{A} ds \int_{-\infty}^{\infty} K(t + \tau + s)\overline{K(t + s)}\, dt = |q(\omega)|^2 \quad (10.24)$$

That is, the sampled spectrum and the true spectrum will have the same time-average value.

For many purposes, we are interested in the approximate spectrum, in which the integration of τ is carried out only over $(0, B)$, where B is 20 seconds in the particular case we have already exhibited. Let us remember that $f(t)$ is real, and that the autocorrelation is a symmetrical function.   Therefore, we can replace integration from 0 to B by integration from $-B$ to B:

$$\int_{-B}^{B} e^{-2\pi i u \tau}\, d\tau \int_{-\infty}^{\infty} dx(t_1, \alpha) \int_{-\infty}^{\infty} dx(t_2, \alpha) \frac{1}{A} \int_{0}^{A} K(t_1 + \tau + s)$$
$$\times \, \overline{K(t_2 + s)}\, ds \quad (10.25)$$

This will have as its mean

$$\int_{-B}^{B} e^{-2\pi i u \tau}\, d\tau \int_{-\infty}^{\infty} K(t + \tau)\overline{K(t)}\, dt = \int_{-B}^{B} e^{-2\pi i u \tau}\, d\tau \int_{-\infty}^{\infty} |q(\omega)|^2 e^{2\pi i \tau \omega}\, d\omega$$

$$= \int_{-\infty}^{\infty} |q(\omega)|^2 \, \frac{\sin 2\pi B(\omega - u)}{\pi(\omega - u)}\, d\omega \quad (10.26)$$

The square of the approximate spectrum taken over $(-B, B)$ will be

$$\left| \int_{-B}^{B} e^{-2\pi i u \tau}\, d\tau \int_{-\infty}^{\infty} dx(t_1, \alpha) \int_{-\infty}^{\infty} dx(t_2, \alpha) \right.$$
$$\left. \frac{1}{A} \int_{0}^{A} K(t_1 + \tau + s)\overline{K(t_2 + s)}\, ds \right|^2$$

which will have as its mean

$$\int_{-B}^{B} e^{-2\pi i u \tau}\, d\tau \int_{-B}^{B} e^{2\pi i u \tau_1}\, d\tau_1 \frac{1}{A^2} \int_{0}^{A} ds \int_{0}^{A} d\sigma \int_{-\infty}^{\infty} dt_1 \int_{-\infty}^{\infty} dt_2$$

$$\times \, [K(t_1 + \tau + s)\overline{K(t_1 + s)}\, \overline{K(t_2 + \tau_1 + \sigma)}K(t_2 + \sigma)$$
$$+ \; K(t_1 + \tau + s)\overline{K(t_2 + s)}\, \overline{K(t_1 + \tau_1 + \sigma)}K(t_2 + \sigma)$$
$$+ \; K(t_1 + \tau + s)\overline{K(t_2 + s)}\, \overline{K(t_2 + \tau_1 + \sigma)}K(t_1 + \sigma)]$$

$$= \left[\int_{-\infty}^{\infty} |q(\omega)|^2 \, \frac{\sin 2\pi B(\omega - u)}{\pi(\omega - u)}\, d\omega \right]^2$$

$$+ \int_{-\infty}^{\infty} |q(\omega_1)|^2\, d\omega_1 \int_{-\infty}^{\infty} |q(\omega_2)|^2\, d\omega_2$$

$$\times \left[\frac{\sin 2\pi B(\omega_1 - u)}{\pi(\omega_1 - u)} \right]^2 \frac{\sin^2 A\pi(\omega_1 - \omega_2)}{\pi^2 A^2 (\omega_1 - \omega_2)^2}$$

$$+ \int_{-\infty}^{\infty} |q(\omega_1)|^2\, d\omega_1 \int_{-\infty}^{\infty} |q(\omega_2)|^2\, d\omega_2$$

$$\times \; \frac{\sin 2\pi B(\omega_1 + u)}{\pi(\omega_1 + u)} \frac{\sin 2\pi B(\omega_2 - u)}{\pi(\omega_2 - u)} \frac{\sin^2 A\pi(\omega_1 - \omega_2)}{\pi^2 A^2 (\omega_1 - \omega_2)^2}$$

$$(10.27)$$

It is well known that, if m is used to express a mean,

$$m[\lambda - m(\lambda)]^2 = m(\lambda^2) - [m(\lambda)]^2 \tag{10.28}$$

Hence the root-mean-square error of the approximate sampled spectrum will be equal to

$$\sqrt{\begin{aligned} & \int_{-\infty}^{\infty} |q(\omega_1)|^2 \, d\omega_1 \int_{-\infty}^{\infty} |q(\omega_2)|^2 \, d\omega_2 \frac{\sin^2 A\pi(\omega_1 - \omega_2)}{\pi^2 A^2(\omega_1 - \omega_2)^2} \\ & \times \left(\frac{\sin^2 2\pi B(\omega_1 - u)}{\pi^2(\omega_1 - u)^2} + \frac{\sin 2\pi B(\omega_1 + u)}{\pi(\omega_1 + u)} \frac{\sin 2\pi B(\omega_2 - u)}{\pi(\omega_2 - u)} \right) \end{aligned}} \tag{10.29}$$

Now,

$$\int_{-\infty}^{\infty} \frac{\sin^2 A\pi u}{\pi^2 A^2 u^2} \, du = \frac{1}{A} \tag{10.30}$$

Thus

$$\int_{-\infty}^{\infty} g(\omega) \frac{\sin^2 A\pi(\omega - u)}{\pi^2 A^2(\omega - u)^2} \, d\omega \tag{10.31}$$

is $1/A$ multiplied by a running weighted average of $g(\omega)$. In case the quantity averaged is nearly constant over the small range $1/A$, which is here a reasonable assumption, we shall obtain as an approximate dominant of the root-mean-square error at any point of the spectrum

$$\sqrt{\frac{2}{A} \int_{-\infty}^{\infty} |q(\omega)|^4 \frac{\sin^2 2\pi B(\omega - u)}{\pi^2(\omega - u)^2} \, d\omega} \tag{10.32}$$

Let us notice that if the approximate sampled spectrum has its maximum at $u = 10$, its value there will be

$$\int_{-\infty}^{\infty} |q(\omega)|^2 \frac{\sin 2\pi B(\omega - 10)}{\pi(\omega - 10)} \, d\omega \tag{10.33}$$

which for smooth $q(\omega)$ will not be far from $|q(10)|^2$. The root-mean-square error of the spectrum referred to this as a unit of measurement will be

$$\sqrt{\frac{2}{A} \int_{-\infty}^{\infty} \left| \frac{q(\omega)}{q(10)} \right|^4 \frac{\sin^2 2\pi B(\omega - 10)}{\pi^2(\omega - 10)^2} \, d\omega} \tag{10.34}$$

and hence no greater than

$$\sqrt{\frac{2}{A} \int_{-\infty}^{\infty} \frac{\sin^2 2\pi B(\omega - 10)}{\pi^2(\omega - 10)^2} \, d\omega} = 2\sqrt{\frac{B}{A}} \tag{10.35}$$

In the case we have considered, this will be

$$2 \sqrt{\frac{20}{2700}} = 2 \sqrt{\frac{1}{135}} \approx \frac{1}{6} \qquad (10.36)$$

If we assume then that the dip phenomenon is real, or even that the sudden fall-off which takes place in our curve at a frequency of about 9.05 cycles per second is real, it is worth while asking several physiological questions concerning it. The three chief questions concern the physiological function of these phenomena which we have observed, the physiological mechanism by which they are produced, and the possible application which can be made of these observations in medicine.

Note that a sharp frequency line is equivalent to an accurate clock. As the brain is in some sense a control and computation apparatus, it is natural to ask whether other forms of control and computation apparatus use clocks. In fact most of them do. Clocks are employed in such apparatus for the purpose of gating. All such apparatus must combine a large number of impulses into single impulses. If these impulses are carried by merely switching the circuit on or off, the timing of the impulses is of small importance and no gating is needed. However, the consequence of this method of carrying impulses is that an entire circuit is occupied until such time as the message is turned off; and this involves putting a large part of the apparatus out of action for an indefinite period. It is thus desirable in a computing or control apparatus that the messages be carried by a combined on-and-off signal. This immediately releases the apparatus for further use. In order for this to take place, the messages must be stored so that they can be released simultaneously, and combined while they are still on the machine. For this a gating is needed, and this gating can be conveniently carried out by the use of a clock.

It is well known that, at least in the case of the longer nerve fibers, nerve impulses are carried by peaks whose form is independent of the manner in which they are produced. The combination of these peaks is a function of the synaptic mechanism. In these synapses, a number of incoming fibers are linked to an outgoing fiber. When the proper combination of incoming fibers fires within a very short interval of time, the outgoing fiber fires. In this combination, the effect of the incoming fibers in certain cases is additive, so that if more than a certain number fire, a threshold is reached which permits the outgoing fiber to fire. In other cases some of the incoming fibers have an inhibitory action, absolutely preventing the firing, or at any rate increasing the threshold for the other fibers. In either case, a short combination period is essential, and if the incoming messages do not lie within this short period, they do not combine. It is therefore necessary to have some sort of gating mechanism to permit the incoming messages to arrive substantially simultaneously. Otherwise the synapse will fail to act as a combining mechanism.[1]

It is desirable, however, to have further evidence that this gating actually takes place. Here some work of Professor Donald B. Lindsley of the psychology department of the University of California at Los Angeles is relevant.   He has made a study of reaction times for visual signals.   As is well known, when a visual signal arrives, the muscular activity which it stimulates does not occur at once, but after a certain delay.   Professor Lindsley has shown that this delay is not constant, but seems to consist of three parts.   One of these parts is of constant length, whereas the other two appear to be uniformly distributed over about 1/10 second.   It is as if the central nervous system could pick up incoming impulses only every 1/10 second, and as if the outgoing impulses to the muscles could arrive from the central nervous system only every 1/10 second.   This is experimental evidence of a gating; and the association of this gating with 1/10 second, which is the approximate period of the central alpha rhythm of the brain, is very probably not fortuitous.

So much for the function of the central alpha rhythm.   Now the question arises concerning the mechanism producing this rhythm. Here we must bring up the fact that the alpha rhythm can be driven by flicker. If a light is flickered into the eye at intervals with a period near 1/10 second, the alpha rhythm of the brain is modified until it has a strong component of the same period as the flicker. Unquestionably the flicker produces an electrical flicker in the retina, and almost certainly in the central nervous system.

There is, however, some direct evidence that a purely electrical flicker may produce an effect similar to that of the visual flicker. This experiment has been carried out in Germany. A room was made with a conducting floor and an insulated conducting metal plate suspended from the ceiling.   Subjects were placed in this room, and the floor and the ceiling were connected to a generator producing

---

[1] This is a simplified picture of what happens, especially in the cortex, since the all-or-none operation of neurons depends on their being of a sufficient length so that the remaking of the form of the incoming impulses in the neuron itself approaches an asymptotic form.   However, in the cortex for example, owing to the shortness of the neurons, the necessity of synchronization still exists, although the details of the process are much more complicated.

an alternating electrical potential which may have been at a frequency near 10 cycles per second. The experienced effect on the subjects was very disturbing, in much the same manner as the effect of a similar flicker is disturbing.

It will, of course, be necessary for these experiments to be repeated under more controlled conditions, and for the simultaneous electroencephalogram of the subjects to be taken. However, as far as the experiments go, there is an indication that the same effect as that of the visual flicker may be generated by an electrical flicker produced by electrostatic induction.

It is important to observe that if the frequency of an oscillator can be changed by impulses of a different frequency, the mechanism must be non-linear. A linear mechanism acting on an oscillation of a given frequency can produce only oscillation of the same frequency, generally with some change of phase and amplitude. This is not true for non-linear mechanisms, which may produce oscillations of frequencies which are the sum and differences of different orders, of the frequency of the oscillator and the frequency of the imposed disturbance. It is quite possible for such a mechanism to displace a frequency; and in the case which we have considered, this displacement will be of the nature of an attraction. It is not too improbable that this attraction will be a long-time or secular phenomenon, and that for short times this system will remain approximately linear.

Consider the possibility that the brain contains a number of oscillators of frequencies of nearly 10 per second, and that within limitations these frequencies can be attracted to one another. Under such circumstances, the frequencies are likely to be pulled together into one or more little clumps, at least in certain regions of the spectrum. The frequencies that are pulled into these clumps will have to be pulled away from somewhere, thus causing gaps in the spectrum, where the power is lower than that which we should otherwise expect. That such a phenomenon may actually take place in the generation of brain waves for the individual whose autocorrelation is shown in Fig. 9 is suggested by the sharp drop in the power for frequencies above 9.0 cycles per second. This could not easily have been discovered with the low resolving powers of harmonic analysis used by earlier writers.[1]

[1] I must say that some evidence of the existence of narrow central rhythms has been obtained by Dr. W. Grey Walter of the Burden Neurological Institute in Bristol, England. I am not acquainted with the full details of his methodology; however, I understand that the phenomenon to which he refers consists in the fact that in his toposcopic pictures of brain waves, as one goes out from the center, the rays indicating the frequency are confined to relatively narrow sectors.

In order that this account of the origin of brain waves should be tenable, we must examine the brain for the existence and nature of the oscillators postulated. Professor Rosenblith of M.I.T. has informed me of the existence of a phenomenon known as the after-discharge.[1] When a flash of light is delivered to the eyes, the potentials of the cerebral cortex which can be correlated with the flash do not return immediately to zero, but go through a sequence of positive and negative phases before they die out. The pattern of this potential can be subjected to a harmonic analysis and is found to have a large amount of power in the neighborhood of 10 cycles. As far as this goes, it is at least not contradictory to the theory of brain wave self-organization that we have given here. The pulling together of these short-time oscillations into a continuing oscillation has been observed in other bodily rhythms, as for example the approximately $23\frac{1}{2}$-hour diurnal rhythm which is observed in many living beings.[2] This rhythm is capable of being pulled into the 24-hour rhythm of day and night by the changes in the external environment. Biologically it is not important whether the natural rhythm of living beings is precisely a 24-hour rhythm, provided it is capable of being attracted into the 24-hour rhythm by the external environment.

An interesting experiment which may throw light on the validity of my hypothesis concerning brain waves could quite possibly be made by the study of fireflies or of other animals such as crickets or frogs which are capable of emitting detectable visual or auditory impulses and also capable of receiving these impulses. It has often been supposed that the fireflies in a tree flash in unison, and this apparent phenomenon has been put down to a human optical illusion. I have heard it stated that in the case of some of the fireflies of Southeastern Asia this phenomenon is so marked that it can scarcely be put down to illusion. Now the firefly has a double action. On the one hand it is an emitter of more or less periodical impulses, and on the other hand it possesses receptors for these impulses. Could not the same supposed phenomenon of the pulling together of frequencies take place? For this work, accurate records of the flashings are necessary which are good enough to subject to an accurate harmonic analysis. Moreover, the fireflies should be subjected to periodic light, as for example from a flashing neon tube, and we should

[1] Barlow, J. S., "Rhythmic Activity Induced by Photic Stimulation in Relation to Intrinsic Alpha Activity of the Brain in Man," *EEG Clin. Neurophysiol.*, **12**, 317–326 (1960).

[2] *Cold Spring Harbor Symposium on Quantitative Biology*, Volume XXV (*Biological Clocks*), The Biological Laboratory, Cold Spring Harbor, L.I., N.Y., 1960.

determine whether this has a tendency to pull them into frequency with itself. If this should be the case, we should try to obtain an accurate record of these spontaneous flashes to subject to an auto-correlation analysis similar to that which we have made in the case of the brain waves. Without daring to pronounce on the outcome of experiments which have not been made, this line of research strikes me as promising and not too difficult.

The phenomenon of the attraction of frequencies also occurs in certain non-living situations. Consider a number of electrical alternators with their frequencies controlled by governors attached to the prime movers. These governors hold the frequencies in comparatively narrow regions. Suppose the outputs of the generators to be combined in parallels on busbars from which the current goes out to the external load, which will in general be subject to more or less random fluctuations due to the turning on and off of light and the like. In order to avoid the human problems of switching which occur in the old-fashioned sort of central station, we shall suppose the switching on and off of the generators to be automatic. When the generator is brought to a speed and phase near enough to that of the other generators of the system, an automatic device will connect it to the busbars, and if by some chance it should depart too far from the proper frequency and phase, a similar device will automatically switch it off. In such a system, a generator which is tending to run too fast and thus to have too high a frequency takes a part of the load which is greater than its normal share, whereas a generator which is running too slow takes a less than normal part of the load. The result is that there is an attraction between the frequencies of the generators. The total generating system acts as if it possessed a virtual governor, more accurate than the governors of the individual governors and constituted by the set of these governors with the mutual electrical interaction of the generators. To this the accurate frequency regulation of electrical generating systems is at least in part due. It is this which makes possible the use of electrical clocks of high accuracy.

I therefore suggest that the output of such systems be studied both experimentally and theoretically in a manner parallel to that in which we have studied the brain waves.

Historically it is interesting that in the early days of alternating-current engineering, attempts were made to connect generators of the same constant-voltage type used in modern generating systems in series rather than in parallel. It was found that the interaction of the individual generators in frequency was a repulsion rather than an attraction. The result was that such systems were impossibly unstable unless the rotating parts of the individual generators were connected rigidly by a common shaft or by gearing. On the other hand the parallel busbar connection of generators proved to have an intrinsic stability which made it possible to unite generators at different stations into a single self-containing system. To use a biological analogy, the parallel system had a better homeostasis than the series system and therefore survived, while the series system eliminated itself by natural selection.

We thus see that a non-linear interaction causing the attraction of frequency can generate a self-organizing system, as it does, for example, in the case of the brain waves we have discussed and in the case of the a-c network. This possibility of self-organization is by no means limited to the very low frequency of these two phenomena. Consider self-organizing systems at the frequency level, say, of infra-red light or radar spectra.

As we have stated before, one of the prime problems of biology is the way in which the capital substances constituting genes or viruses, or possibly specific substances producing cancer, reproduce themselves out of materials devoid of this specificity, such as a mixture of amino and nucleic acids. The usual explanation given is that one molecule of these substances acts as a template according to which the constituent's smaller molecules lay themselves down and unite into a similar macromolecule. This is largely a figure of speech and is merely another way of describing the fundamental phenomenon of life, which is that other macromolecules are formed in the image of the existing macromolecules. However this process occurs, it is a dynamic process and involves forces or their equivalent. An entirely possible way of describing such forces is that the active bearer of the specificity of a molecule may lie in the frequency pattern of its molecular radiation, an important part of which may lie in infra-red electromagnetic frequency or even lower. It may be that specific virus substances under some circumstances emit infra-red oscillations which have the power of favoring the formation of other molecules of the virus from an indifferent magma of amino acids and nucleic acids. It is quite possible that this phenomenon may be regarded as a sort of attractive interaction of frequency. As this whole matter is still *sub judice*, with the details not even formulated, I forbear to be more specific. The obvious way of investigating this is to study the absorption and emission spectra of a massive quantity of virus material, such as the crystal of the tobacco mosaic virus, and then to observe the effects of light of these frequencies on the production of more virus from existing virus in the proper nutrient material. When I speak of absorption spectra, I am talking of a phenomenon which is almost certain to exist; and as to emission spectra, we have something of the sort in the phenomenon of fluorescence.

Any such research will involve a highly accurate method for the detailed examination of spectra in the presence of what would ordinarily be considered excessive amounts of light of a continuous spectrum. We have already seen that we are faced with a similar problem in the microanalysis of brain waves, and that the mathematics of interferometer spectrography is essentially the same as that which we have undertaken here. I then make the definite suggestion that the full power of this method be explored in the study of molecular spectra, and in particular in the study of such spectra of viruses, genes, and cancer. It is premature to predict the entire value of these methods both in pure biological research and in medicine, but I have great hopes that they may be proved to be of the utmost value in both fields.

Index

CPSIA information can be obtained
at www.ICGtesting.com
Printed in the USA
BVHW03s2238250218
508724BV00004B/216/P

9 781614 275022